Lecture Notes in Earth Sciences 79

Editors:
S. Bhattacharji, Brooklyn
G. M. Friedman, Brooklyn and Troy
H. J. Neugebauer, Bonn
A. Seilacher, Tuebingen and Yale

W0232021

Springer-Verlag Berlin Heidelberg GmbH

Geoffrey H. Dutton

A Hierarchical Coordinate System for Geoprocessing and Cartography

With 85 Figures, 16 Tables and 2 Foldouts

 Springer

Author

Geoffrey H. Dutton
Department of Geography, Spatial Data Handling Division
University of Zürich-Irchel
Winterthurerstrasse 190, CH-8057 Zürich/Switzerland
E-mail: dutton@geo.unizh.ch

"For all Lecture Notes in Earth Sciences published till now please see final pages of
the book"

Cataloging-in-Publication data applied for

Die Deutsche Bibliothek - CIP-Einheitsaufnahme

Dutton, Geoffrey H.:
A hierarchical coordinate system for geoprocessing and cartography /
Geoffrey H. Dutton.
  (Lecture notes in earth sciences ; 79)
  ISBN 978-3-540-64980-9   ISBN 978-3-540-49802-5 (eBook)
  DOI 10.1007/978-3-540-49802-5

Additional material to this book can be downloaded from http://extras.springer.com

This monograph was a Ph.D. dissertation approved by the Philosophical Faculty II
of the University of Zürich under the title "Working Through the Scales: A Hierarchical Coordinate System for Digital Map Data".

Die vorliegende Arbeit wurde von der Philosophischen Fakultät II der Universität
Zürich im Wintersemester 1997/98 aufgrund der Gutachten von Prof. Dr. Robert
Weibel und Prof. Dr. Kurt Brassel als Inaugural-Dissertation angenommen.

ISSN 0930-0317

Typesetting: Camera ready by author
SPIN: 10685802   32/3142-543210 - Printed on acid-free paper

For *William Warntz*
Buckminster Fuller
Benoit Mandelbrot

Everything the same; everything distinct
— Zen proverb

Preface

Working on a Ph.D. can be lonely, frustrating, discouraging. In my own case, the work was both made easier and more difficult by having been involved in researching the subject matter of my dissertation for more than fifteen years: easier because I had worked out many of the basic concepts and techniques by the time I began the most recent phase; harder because there was by then a lot of material to integrate, and my main application area (map generalization) has so many complexities and vexing unsolved problems, some of which I attempted to tackle head-on. While I usually managed to budge them, the problems didn't always yield.

The main reason for my good fortune in being able to bring this project to some sort of closure is my connection to vital communities of extraordinary people, who if they did not actively encourage and support my efforts, gracefully tolerated my obscure and idiosyncratic visions with good humor, and possibly even respect. And during my time in Switzerland, I have been able to work steadily and uninterrupted, refreshed by periodic reunions with my family back in America. The unconditional support of my wife, Aygül Balcioglu Dutton and my mother, Sophie Pincus Dutton have nourished my spirit throughout my expatriation. I owe both of them so much.

There are so many others who have helped, in many ways. While this work has been a very personal odyssey, others have collaborated with me from time to time and supported the project. I am especially grateful to the National Center for Geographic Information and Analysis (NCGIA) for encouraging this work, by including me in specialist meetings for Research Initiatives 1, 3, 7 and 8, which prompted me to explore interesting avenues related to this project and resulted in several publications. In particular, I wish to thank Michael Goodchild (Santa Barbara), Barbara Buttenfield (Buffalo, now at the University of Colorado), and Kate Beard (Orono) for their efforts. It was Mike who encouraged Yang Shiren and others to develop the first working implementation of my model, brought us together to work on it, and reported on our progress in several venues. Dr. Shiren was a fountainhead of clever, practical ideas for implementing my data model, and I am highly appreciative for his collaboration. The resulting software prototypes demonstrated to me possibilities I had not considered, and convinced me of the essential value and practicality of these concepts. I am equally indebted to Barbara Buttenfield for inviting me to NCGIA Buffalo on several occasions to collaborate; there the central concepts that led to this project were distilled and the map generalization algorithms were first developed, some of which remain at the core of the current project.

All of the above would have come to naught had it not been for my advisors in Zürich, Kurt Brassel and Robert Weibel. Together they did managed the logistics to overcome various obstacles in my path, and have been extremely supportive and encouraging to me all the way. While in residence, my principal advisor has been Dr. Weibel, whose keen and detailed criticism and useful suggestions have helped to keep me on the right track. I am also grateful to the Swiss National Science Foundation for financial support for this research, through grants 2000-037649.93 and 2100-43502.95 to the Department of Geography. I am indebted to my colleague Frank Brazile as well; his cheerful skepticism prevented me from getting carried away too far with each new putative insight, and he also has brought to my attention tools that have been useful and could become more so as my project grows and matures. Frank's help in computing spatial orderings presented in appendix E is warmly appreciated.

Day in and day out my department proved to be a friendly atmosphere and close to an ideal environment in which to work: I was provided with a workstation, a computer account, access to libraries and software packages, and basically left alone to follow my research agenda. There has been no coursework or exams to distract me, just occasional interesting colloquia, a few administrative chores and the task at hand. Having seized this opportunity, I hope that this and other results of my efforts justify my advisors' and my family's faith in my capabilities.

Still, I have not managed to accomplish everything I had hoped for when this project commenced two years ago. My goals then — and now — included developing a more comprehensive set of tools to solve contextual map generalization problems, and while a good amount of conceptual progress was made toward this end, its implementation is far from complete, indeed barely begun. But this has been a work in progress for close to fifteen years now, and there is no reason for me to expect a sudden dénouement. It is my fond hope that others will find something of value in my work, and will take up the challenge of making it more operational and useful in an increasing variety of contexts. If I can help any of these efforts move forward, it would be my pleasure to collaborate.

Especially, I would like to thank Luisa Tonarelli for her patience, good humor and expertise in guiding my efforts to publish my dissertation in this series.

Geoffrey Dutton   June 1998
Zürich   QTM ID: 1133013130312011310

Abstract

The work reported in this thesis has sought to determine if a scale-specific notation for location can benefit GIS and cartography. After discussing some of the problems caused by relying on standard coordinate notations, the quaternary triangular mesh (QTM), a hierarchical coordinate system based on a planetary polyhedral tessellation is presented. This model encodes terrestrial locations as leaf nodes in a forest of eight triangular quadtrees, each of which can have up to 29 levels of detail (doublings of scale). Its multi-resolution properties are described in relation to common planar and spherical coordinate systems used for geographic location coding. One of QTM's advantages derives from its inherent ability to document the locational accuracy of every point in a spatial dataset. Another is its ability to retrieve such data across a range of spatial resolutions, enabling users to evaluate the fitness of map data for use at particular scales. In addition, the tessellation's spatial partitions can be exploited to identify regions where crowding of features occurs at certain scales. After describing this system's origins and comparing it to related data models, techniques for encoding and decoding locations, accuracy assessment, spatial conflict detection and other operations are presented. The focus then moves to the use of QTM to generalize digital map data for display at smaller scales. After a brief overview of current approaches to generalization, a strategy for using QTM for line simplification and conflict detection is outlined. A basic strategy for using QTM for line simplification was developed that is embodied in an algorithm having eight control parameters. This method is capable of using embedded spatial metadata to guide selection of retained points, and a way to encode such metadata is described; exploitation of point-specific metadata was seen as one of the more successful and promising aspects of the strategy. A series of empirical tests of this algorithm using several island shoreline datasets is described and illustrated. Outcomes of these experiments are compared with results produced by an existing, widely-used line simplification algorithm. Similarities and differences in results are discussed, attempting to identify reasons for failure or success of either method. The thesis concludes with an evaluation of QTM's potential contributions to this application area, along with a description of some of the current challenges and limitations to spatial data handling that QTM may be able to address. Additional overviews of contents may be found by consulting Scope of Work in chapter 1, as well as the summaries at the end of each chapter.

Zusammenfassung

Die vorliegende Arbeit geht der Frage nach, ob eine neue, massstabsunabhängige
Notation für Positionen (Koordinaten) in Geographischen Informationssystemen
(GIS) und in der Kartographie von Vorteil sein könnte. Zunächst werden einige der
Probleme herkömmlicher Korrdinatennotationen untersucht, um dann das
"Quaternary Triangular Mesh (QTM)" als Alternative zur Beschreibung und
Speicherung räumlicher Daten einzuführen. QTM ist ein hierarchisches
Koordinatensystem, das auf einer planetaren polyedrischen Tessellation basiert.
Ausgehend von einem Oktaeder, beschreibt es Positionen auf der Erdkugel als
Blätter in einem Wald von acht Dreiecks-Quadtrees (entsprechend den acht
Oktaederflächen). Jeder Quadtree kann bis zu 29 Auflösungsstufen umfassen (je
Stufe verdoppelt sich der Massstab). Die besonderen Eigenschaften dieser
hierarchischen Datenstruktur, insbesondere die Möglichkeit mehrstufige
Auflösungen innerhalb einer Datenstruktur zu repräsentieren, werden in Relation
zu herkömmlichen planaren und sphärischen Koordinatensystemen beschrieben.
Ein wesentlicher Vorteil dieses Datenmodells ist, durch seine Struktur die
Genauigkeit eines jeden Datenpunktes beschreiben zu können; ein weiterer, dass
man Daten für verschiedene Auflösungen extrahieren und so auf einfache Weise die
Tauglichkeit von Daten für verschiedene Kartenmassstäbe evaluieren kann.
Ausserdem eignet sich die mehrstufige Raumaufteilung gut, um Gebiete mit
zuviel Detail in einem bestimmten Massstabsbereich zu erkennen. Nach der
Diskussion der Ursprünge des QTM-Modells und dem Vergleich mit verwandten
Systemen beschreibt die Arbeit mehrere Algorithmen zur Kodierung und
Dekodierung von Positionen, Genauigkeitsabschätzung, Ermittlung räumlicher
Konflikte (für die kartographische Generalisierung), sowie weitere
Basisoperationen. Danach wendet sich die Arbeit vornehmlich der Verwendung von
QTM zur digitalen Generalisierung kartographischer Daten zu. Nach einer kurzen
Diskussion bestehender Ansätze für die Generalisierung wird eine Strategie für die
Verwendung von QTM für die Linienvereinfachung und für die Erkennung
potentieller Konflikte zwischen kartographischen Objekten vorgestellt. Unter
anderem wurde eine Methode für die QTM-basierte Linienvereinfachung entwickelt,
die zu einem Algorithmus geführt hat, der von acht Parametern gesteuert wird.
Dieser Generalisierungsalgorithmus kann durch Metadaten, die im QTM abgelegt
werden, gezielt gesteuert werden und so sicherstellen, dass wesentliche Punkte
einer Linie erhalten bleiben. Eine Methode für die Extraktion von Metadaten zur
Beschreibung der Sinuosität kartographischer Linien und deren Kodierung im QTM
wird ebenfalls beschrieben. Die Verwendung von punktbezogenen Metadaten zur

Steuerung der Generalisierungsalgorithmen hat sich als sehr vorteilhafter und vielversprechender Aspekt der vorliegenden Strategie erwiesen. Dies wird auch durch eine Reihe empirischer Tests erhellt, die anhand verschiedener Datensätze von Küstenlinien durchgeführt wurden. Die Ergebnisse dieser Untersuchungen werden einem herkömmlichen, weit verbreiteten Linienvereinfachungsalgorithmus gegenübergestellt. Ähnlichkeiten und Unterschiede in den Resultaten der zwei Ansätze werden diskutiert, um so die Gründe für Erfolg oder Misserfolg der jeweiligen Methode zu identifizieren. Die Arbeit schliesst mit einer Diskussion konkreter Anwendungsbereiche für QTM und seinem potentiellen Beitrag zur Lösung aktueller Probleme Geographischer Informationssysteme und in der Kartographie. Weitere Hinweise über den Inhalt der vorliegenden Arbeit können dem Kapitel 1 (Abschnitt Scope of Work) sowie den Kurzzusammenfassungen am Ende jedes Kapitels entnommen werden.

Contents

List of Figures

List of Tables

Abbreviations and Acronyms

AID	attractor identifier
CSL	classified sinuosity, local
CSR	classified sinuosity, regional
CSV	classified sinuosity value
DBMS	database management system
DEM	digital elevation model
DEPTH	delta-encoded polynomial terrain hierarchy

DLG	digital line graph
DTM	digital terrain model
EMAP	Environmental Monitoring and Assessment Program
EPA	(U.S.) Environmental Protection Agency
FGDC	(U.S.) Federal Geographic Data Committee
FID	Feature Identifier
GEM	geodesic elevation model
GIS	geographic information system
HSDS	hierarchical spatial data structure
MBT	minimum bounding triangle
MEL	mesh element
NMAS	national map accuracy standard
NOAA	(U.S.) National Ocean and Atmospheric Agency
PID	primitive identifier
PSV	preferred sinuosity value
QTM	quaternary triangular mesh
RDP	Ramer-Douglas-Peucker (algorithm)
SDTS	(U.S.) Spatial Data Transfer Standard
SQT	sphere quad tree
TIN	triangulated irregular network
TQS	triangulated quadtree surface
USGS	United States Geological Survey
UTM	universal transverse Mercator
WWW	world wide web
ZOT	zenithial ortho triangular (projection)

1 Introduction and Problem Statement

One real world is enough.
— Santayana

When a mapping agency updates information on topographic maps or navigational charts, usually more than one map is affected by a given change. Often, the largest (most detailed) scale maps are edited first, and the changes then transferred to the appropriate map at the next smaller scale, until either the smallest scale is reached or the edited feature becomes too small to represent. As a result of changes in the world — or interpretations of it — marks on maps may change their shape or symbology, relative prominence, locations and labels, according to organizational standards and procedures as well as expert judgement, also known as "cartographic license." This tedious graphic editing process has been described as "working through the scales," as it systematically propagates edits from greater to less detailed cartographic representations. It is one strategy of attacking problems in the art of *map generalization*, one which specialists from many nations are struggling to understand, formalize and automate.

Until about a decade ago, map generalization was practiced almost exclusively by trained cartographers, who learned it by tutelage, example and intuition, deliberately, even rigorously, but not necessarily formally. Manual cartographic generalization involves an unruly set of techniques and precepts that, traditionally, were mainly practiced in public agencies and map publishing houses, and each organization tended to have its own procedures, standards, guidelines and aesthetics for compiling maps in fulfilment of its mission. Only recently has this situation begun to change, but it has done so dramatically and definitively, as maps — along with every other form of human communication — have become digital. At the same time, the media, data, tools and enterprises involved in compiling and communicating maps are rapidly becoming global. The author of this thesis, in describing, demonstrating and assessing a specific digital approach to working through scales, hopes to contribute to progress in both map generalization specifically and spatial data handling generally.

1.1 Motivation

While the foregoing describes its application focus, this thesis has a deeper, more general concern: improving the representation of geographic space — the world at large — in digital computers. The latter topic is necessary to consider because addressing the former one will achieve limited success without re-thinking how encoding conventions for geospatial data bias the development and constrain the results of methods used to generalize it. This may not appear obvious or even necessary to many who work with environmental, cartographic and other spatial datasets on a regular basis, as the software technology for handling them seems to be relatively mature and generally effective. Furthermore, a number of standards for specifying and exchanging such data are in use; these may differ in detail, but generally assume certain encoding conventions for basic map elements that are rarely questioned anymore. As a consequence, this thesis argues, progress in map generalization, spatial data interoperability and analytic applications has been hampered due to encountering a number of vexing difficulties, many of which stem from inadequate descriptions of both geographic phenomena and the space they inhabit.

1.2 Limitations to Detail and Accuracy of Spatial Data

We start by confronting a paradox: when geographic phenomena that are spatially continuous — such as terrain, temperature or rainfall — are numerically modeled, an inherently *discontinuous data structure* (regular sampling *grids*, or *rasters*, sometimes called *fields*)[1] is often employed. Yet when the geometry of inherently discontinuous, linear or sharp-edged phenomena — such as administrative boundaries, land ownership, hydrography or transportation — is encoded, it tends to be done using a *continuous* data structure capable of representing the shapes and connections between entities and their neighbors in great detail (so-called *vector-topological* models). Some highly interpreted geospatial data, such as land use, soils, geologic structures, classified remote-sensing imagery and results of migration and accessibility models may be represented in either vector or raster format, although making either choice requires certain compromises.

If they have not worked in the raster domain, many users of geographic information systems (*GIS*) may not be aware that digital map data they use are not as definitive and concrete as they may appear to be. Those who work with raster-based GIS and image processing (IP) systems, which store and manipulate images and data grids, need little reminding of the finite resolution of their databases. Each grid *cell* and pixel in every dataset they handle covers a specific amount of territory on, above or under the surface of the earth, and cannot disclose details of variations within these tiny but specific domains. And raster *data models* — even multi-resolution schemes such as *quadtrees* — cannot hide the fact that only a certain amount of detail is stored, amounting to very many discrete (but not independent) observations of single-valued attributes.

---

1 Terms defined in the glossary, appendix B, are printed in bold italics where they first appear.

In non-raster GIS environments — with which this project is almost exclusively concerned — the discrete nature of data and the limits of its *precision* and *accuracy* tend to be hidden, and may go unnoticed, but not without consequence. Systems that employ vector-topological data models actually introduce two types of discretization, illustrated in fig. 1.1:

1 Continuous contours of spatial entities are represented by finite, ordered sets of *vertex* locations — although sometimes, continuous functions (such as splines or elliptical arcs) may be exploited to represent certain types of curves. The *accuracy* of such data are mainly limited by sampling during their capture;

2 Based on how much computer storage is assigned to primitive data types (integers and real numbers, which in practice usually is either 32 or 64 bits), the *precision* of individual spatial *coordinates* (vertex locations) is always finite.

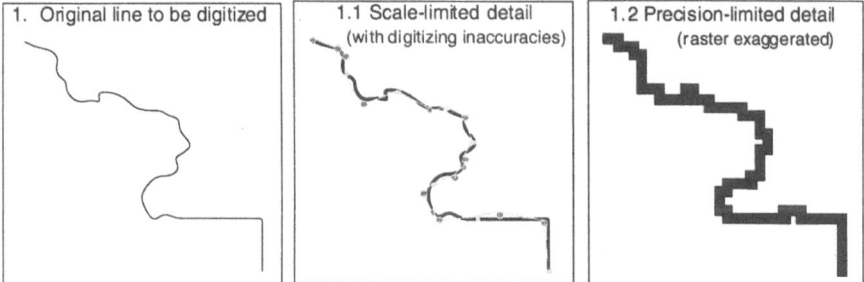

Fig. 1.1. Two types of discretization errors, along with human line tracing error

1.2.1 Scale Limitations

The first type, which we will denote as *scale-limited*[2] representation, caricatures *lines* and polygonizes *regions* by sampling networks and boundaries at small (and usually irregular) intervals. The size of intervals is mainly influenced by data capture technology and procedures and the scale and resolution of the graphic source material, which typically are aerial photographs or printed maps, captured as digital data (digitized) either automatically or manually. If assumptions are made

---

[2] In this document, *scale* is synonymous with *map scale*; a dimensionless ratio of distance on a map to the "true" distance that it represents on the earth's surface. Such ratios are always less than unity, and are notated as 1:N, where N is usually between 100 and 10,000,000. When N alone is referred to, it is termed the *scale denominator*.

regarding the smallest mark or object that a map or a display is able to represent, one may numerically relate spatial accuracy and resolution to *map scale* (Tobler 1988, Goodchild 1991a, Dutton 1996). The spatial data model explored in this thesis makes use of such equivalencies (see table 2.1).

GIS users and cartographers are well-conditioned by now to line segment approximations of map features, and have some understanding of their limitations and of consequences to merging or overlaying digital map data. GIS system designers have come up with a variety of solutions to the "sliver problem," the generation of small, spurious shapes resulting from mismatches of line-work that represents the same features using slightly different geometric descriptions. Paradoxically, as Goodchild (1978) has shown, the more diligent and detailed is one's digitization effort, the more troublesome this problem gets. No practical, robust GIS solutions are yet available for automatically changing the scale of digital map data, or combining data from maps having major differences in the scales at which their features were captured. This is partially due to the fact that GISs — even when metadata are available to them — do not formally model the information content of the data they handle, even the critical yet restricted aspects of *data quality* involving spatial resolution, accuracy and positional error. In a Ph.D. thesis concerned with handling scale- and resolution-limited GIS data, Bruegger proposes an approach to data integration in which:

> ... mapping a representation to a coarser resolution becomes well-defined since the source and target knowledge content are precisely known. This compares to current GIS practice where the knowledge content of a representation is only vaguely known to the user and totally inaccessible to machine interpretation. Figure [1.2] illustrates this with an example of two vector representations of the same coast line. What knowledge about the world do the representations really contain? For example, what is found in location p? The vague definition of the knowledge content is closely related to the problem of precisely defining what it means to transform a representation from one scale to another.
>
> (Bruegger 1994: 4)

Bruegger's goal was to develop a "format-independent" representation to enable conversion between formats (specifically, raster and vector) to take place without unnecessary loss of information. Such a notation could specify the positional uncertainty of the geometry contained in a dataset as metadata equally applicable to each type of representation. This, he asserted, would avoid "problems of comparing incompatible concepts such as raster resolution and vector 'scale'" (Bruegger 1994: 4).

The latter statement is arguable, we maintain; there is nothing incompatible between these two concepts if one makes certain assumptions about what they mean in an operational sense, as they both are consequences of discretization processes, one with respect to space and the other with respect to phenomena. This aside, the goals of this thesis are in harmony with Bruegger's, even though its approach, formalisms and application concerns are somewhat different.

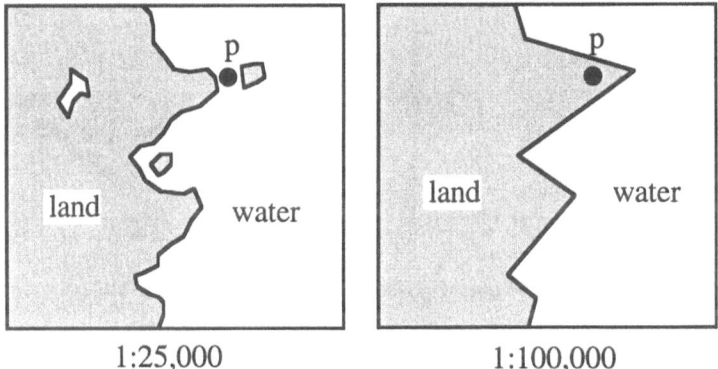

1:25,000   1:100,000

Fig. 1.2. Scale-limited representational problems (from Bruegger 1994: 4)

1.2.2 Precision Limitations

We call the second type of discretization *precision-limited*[3], as it is constrained by how many digits of precision computer hardware or software can generate or represent. This applies to data capture hardware (e.g., scanners and digitizing tables), which have finite-precision grids through which **points** are filtered, as well as to the word sizes supported by a computer's CPU (central processing unit) or implemented in software by programming languages. The basic issue is that geometric coordinates are often modeled as real numbers, but become represented by finite, floating point quantities that can fail to distinguish data points that are very close together and which can introduce numerical errors and instabilities, hence cause algorithms to make incorrect decisions when testing and comparing computed values. A good example of how geometric imprecision can lead to topological trouble is described and illustrated by Franklin (1984). In figure 1.3 scaling is applied to a triangle and a point, resulting in the movement of the point from inside to outside the triangle. The figure exaggerates the effect by using integer coordinates, but the principle holds true for floating point ones as well.

---

[3] Following Goodchild (1991), we define *precision* as "the degree of detail available for a measurement". Note that effective precision may be less than the theoretical maximum, due to rounding, cumulative processing errors and (in floating-point words) normalization of mantissas.

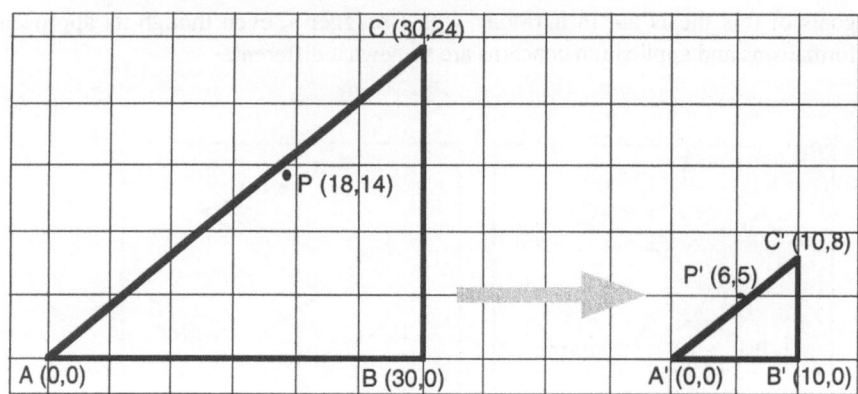

Fig. 1.3. Scaling moves point P outside triangle (from Franklin 1984)

Franklin's paper provides other sobering lessons in the hazards of computational geometry, and provides a look at some ways to avoid them. Although numerical methods have improved considerably since he wrote, his conclusions remain valid:

> Existing computer numbers violate all the axioms of the number systems they are designed to model, with the resulting geometric inaccuracies causing topological errors wherein the database violates its design specifications. (Franklin 1984: 206)

It has found to be devilishly difficult to guarantee the robustness of algorithms that process geometric data, at least unless limiting assumptions are made about the nature and quality of input data, or should approximate answers suffice. Considerable effort in computer graphics and computational geometry has been devoted to handling the consequences of finite-precision computations in processing geometric data (Schirra 1997). While these issues are relatively well-understood, and robust techniques for dealing with some of them exist, this is not the end of the story where processing geospatial data is concerned. One reason is that many of the techniques for avoiding errors in geometric computations utilize software representation of numbers that extend their intrinsic precision generally, or as required by specific computations. While they give more accurate results than hardware floating point computations, they are hundreds or thousands of times slower to execute. This may be tolerable for some kinds of geometric applications, but GIS applications tend to be too data-intensive to make such approaches practical.

Also, even if GIS processing modules could be made provably correct in handling all the precision available for map and other locational data, its degree tends to be fixed, and is rarely sensitive to variations in the actual precision of coordinates within datasets. Finite precision is basically regarded as a nuisance, to be overcome by force, usually by allocating more bits than will ever be needed to store and process coordinate data (e.g., double-precision floating point words). This almost always fails to reflect the true amount of information that a coordinate contains about a *geographic* location, tending to exaggerate it.

Standardizing (hence falsifying) the precision of locations to machine word sizes may result in false confidence in the quality of computations performed on them, even if all numerical instabilities are carefully avoided. It may lead one to believe, for example, that a boundary location is known to within 2 cm, when in fact the measurement is at best accurate to about 20 meters. Some modules of some GISs do provide ways to simulate reductions to precision (by storing accuracy parameters and introducing error tolerances); these, however, are almost always based on global estimates or information about measurements (error *attributes* or *metadata*), rather on the actual precision of measurements themselves. Another common strategy is to employ rational arithmetic, using integral numerators and denominators to represent real numbers. While this can ensure unambiguous results for computations such as line intersection or point-in-polygon tests, the results tend to be only as good as the least accurate coordinate involved.

1.3 Documenting Spatial Data Quality

But data — spatial and otherwise — are not only imprecise, they also (always!) contain *errors* and *inaccuracies*. The caricaturing of boundaries described above in discussing types of discretization is fundamentally a lack of accuracy[4]. Inaccuracy is inevitable, and up to a point acceptable, given that all data are abstractions of reality that by their nature must leave out details that either escape detection or would serve no useful purpose to include. If effective quality control is maintained during data capture and processing, it is normally possible to provide users with reasonable estimates of the types and magnitudes of inaccuracy within datasets. By properly interpreting such statistics, users (and software) should in principle be able to decide if a given dataset has sufficient fidelity to reality to serve a particular purpose. This is a basic tenet of the U.S. Spatial Data Transfer Standard (SDTS, USDOC 1992), and a primary impetus to the generation and cataloguing of spatial metadata, now becoming a light industry of sorts as geospatial data pops up on networked file servers around the globe.

While spatial data may be inaccurate in a variety of respects, this thesis is principally concerned with *positional accuracy*, and in particular, with how it relates to efforts to generalize topographic maps. Here is a specification for positional accuracy on map sheets published by the United States Geological Survey (USGS), which is responsible for compiling topographic maps for the U.S. at a series of scales from 1:24,000 down:

---

[4] Again, using Goodchild's definition (Goodchild 1991), *accuracy* is the degree to which measurements faithfully represent the real-world entities they purport to describe.

The United States National Map Accuracy Standard (NMAS) specifies that 90% of the well-defined points that are tested must fall within a specified tolerance. For map scales larger than 1:20,000, the NMAS tolerance is 1/30 inch (0.85 mm), measured at publication scale. For map scales of 1:20,000 or smaller, the NMAS tolerance is 1/50 inch (0.51 mm), measured at publication scale.

Converting to ground units, NMAS accuracy is:

S / 360 feet = (1/30 inch) * (1 ft/12 in)*S,
  for map scales larger than 1:20,000 (= S / 1181 m)
S / 600 feet = (1/50 inch) * (1 ft/12 in)*S,
  for map scales of 1:20,000 or smaller (= S / 1969 m)

where S is the map scale denominator.   (FGDC 1996)

Note that this statement only defines horizontal accuracy for "well-defined points," locations which are visible and identifiable in the field. Consequently, it may be difficult to assess the accuracies of map features such as streams, political boundaries, pipelines, contour lines or even roadways, unless they include monumented points. Spatial metadata (FGDC 1994) may give spatial data users some hints about how well such features reflect reality, but the information may be qualitative or narrative, not readily usable for GIS processing purposes. In addition, as most metadata refer to an entire dataset, differentiation of data quality between or within feature classes is generally not possible unless it is specified via attribute coding.

The U.S. NMAS was defined in 1947, well before digital mapping evolved from a curiosity in the 1960's to its status as an industry today. It is widely recognized that this standard is barely adequate for printed maps, and leaves much to be desired in the realm of digital map data. It is in the process of being updated by the U.S. Federal Geographic Data Committee (FGDC 1996); this will not change the basic approach, which may be practical for map production organizations, but is not especially helpful in certifying or assessing accuracies of distributed, heterogeneous spatial datasets that are increasingly created by and available to communities of GIS users. As a result, positional accuracy of most spatial datasets remains murky, and must often be assumed, guessed, or estimated. Sometimes this can be done using existing GIS functions, such as Goodchild and Hunter (1997) demonstrate, but many users would rather not know or will not take the trouble.

Even when the highest data capture standards are followed, and even when codebooks and metadata are scrupulously prepared to document datasets, the fact remains that geospatial data do not describe their internal quality variations very well, if at all. The reasons why this problem persists are many, and can be categorized as:

1 **Circumstantial**: Source data are not well controlled or documented, or are too diverse or convolved to compile data quality descriptions for them;

2 **Institutional**: Datasets are prepared for specific, limited or internal purposes, without a mandate to inform other potential data users;

3 **Structural**: No adequate mechanisms are in general use which are capable of documenting variations in spatial data quality at a highly detailed level.

The first two of these three categories are being dealt with in the GIS community by deliberate efforts, mostly through developing richer spatial data exchange formats and standards for geospatial metadata. As both public and private-sector data producers seek to add value to map data products and make them available over wide-area networks, particularly the internet, data documentation and quality control is receiving a great deal of recent and salutary attention. It is already possible to browse metadata repositories distributed throughout the planet on the world wide web (WWW), to identify (and in many cases also download) geospatial datasets with reasonable certainty about the nature of their contents.[5] But even with the best of intentions, efforts and tools, it remains quite problematic to assess the suitability of *geodata* for purposes and scales other than those for which their producers intended them to be used.

It is argued here that many limitations to reusability of geospatial data as well as many difficulties involved in their maintenance, are due to the third aspect of documenting their quality: structural deficiencies in datatypes and data models. The most glaring, yet largely unacknowledged, deficiency of GIS and cartographic vector data is its reliance on coordinate notation — (latitude, longitude, elevation) or (x, y, z) — to describe locations. This convention is so well-established and ingrained that it is hardly ever questioned, but without doubt it is responsible for millions of hours of computation time and billions of geometric errors that might have been avoided had richer notations for location been used. The author has articulated this previously (Dutton 1984; 1989; 1989a; 1992; 1996).

Regardless of how many dimensions or digits a conventional coordinate tuple contains, it is descriptive only of position, and does not convey scale, accuracy or specific role. It is possible to convey a sense of accuracy by varying the number of digits that are regarded as significant in a coordinate, but such a device rarely is used, but never to distinguish one boundary point from the next (Dutton 1992). In most digital map databases, the vast majority of coordinates are convenient fictions, as few of them represent "well-known points" on the surface of the Earth. Rather than identifying specific points on the Earth's surface, most map coordinates should be considered as loci of events that led to their creation. These

---

5 In the U.S., any internet site that serves spatial data can be registered as a node in the National Spatial Data Infrastructure, by documenting its holdings in compliance with the FGDC's Content Standards for Digital Geospatial Metadata (FGDC 1994). The FGDC, academic organizations and GIS vendors are making available tools which ease the burden of compiling, cataloging and searching such documents.

events are partly natural (geological changes, for example), but may also be of human origin (such as territorial claims), and include activities involved in data capture and processing. Despite growing attention to error in spatial data (Goodchild and Gopal 1989; Veregin 1989; Guptill and Morrison 1995) spatial analysts, cartographers and their software tend to treat coordinates as if they have physical existence, like protons or pebbles. Most vector data structures tend to reify and democratize feature coordinates (although endpoints are usually given special node status). When processing boundaries, most applications treat a tuple that represents a specific location (such as monuments, corners and posts) the same way as a less well-defined one (such as inflections along soil boundaries, roadways and river courses), as just another point, or just another node. Their data structures have no way to express variations in positional data quality, and not surprisingly, their algorithms have no way to use such information. It is a vicious cycle, entrenching ignorance.

One could also call this attitude toward data quality the *fallacy of coordinates* (Dutton 1989), and is an example of the more general *fallacy of misplaced concreteness* ("if the computer said it, then it must be true"). The question for spatial data is, how can we tell if it's true, or more specifically, how true it might be?

1.3.1 Data Quality Information and Map Generalization

What connections might metadata and map generalization have? It is clear to a number of researchers (Mark 1989, Mark 1991, McMaster 1991) that the more information available to describe the nature of map features and their roles in a landscape, the more intelligently it is possible to treat them when changing the scale of a map or creating a map for a specific purpose. Some of this information might appear as tabular attributes to map features, or as global descriptors to specialized datasets. Knowing that a hydrographic feature is, for example, one bank of a braided stream can inform a generalization process applied to it, and potentially modify its behavior compared to how it would handle an ordinary stream centerline representing a main channel. Here is an example of specifications for coding and digitizing braided streams in the Digital Line Graph (*DLG*) format, taken from internal USGS documentation (USGS 1994):

050 0413 Braided stream
This code identifies braided streams that are shown by symbols 404(A), 541.6, 541.9 (C), or 2202.03(D). A braided stream is a special case where the stream subdivides into interlacing channels. In map compilation, where possible, the actual channels are shown. However, if the channels are extremely complex or obscured by vegetation, the outer limit is scribed accurately and the inner channels are represented by a conventional pattern. The use of pattern versus actual channel is not noted on the map. Therefore, the braided portion of a stream is digitized as an area that carries this code. The outer limits are digitized and carry left and right bank

codes (see codes 050 0605 and 050 0606). The braided area is separated from a double-line stream by a closure line (code 050 0202) and from a single-line stream by nodes (see codes 050 0004 and 050 0005).

This USGS document takes nearly 250 pages (of which the section on hydrography accounts for 40) to describe just the attribute codes, associated symbol codes and instructions such as the above. While it contains a considerable amount of knowledge about map features and what they represent, it does not include specific advice on how, where or when features should be generalized. The amount of detail in the DLG coding guide is impractical to provide as file-specific metadata, but it could be turned into a knowledge base (KB), formalized as production rules or via other schemata, given sufficient effort and motivation. Once such a KB for digital database production is built, additional rules for generalization can be added incrementally, to the extent they can be derived from formal descriptions or actual cartographic practice.

As the above DLG guideline states, precisely locating a stream bank may not be feasible in places where vegetation obscures it. It is quite common for positional uncertainty of boundaries to change along a given feature for this and a variety of other reasons, such as construction or other earth moving activities, ecological succession, the presence of wetlands, and when map data from different sources is merged in developing a GIS database. When the level of uncertainty of a feature (or portion of one) changes from the norm for its data layer or feature class, most GISs — although they could — tend not record this, as it requires adding error attributes that will only be occasionally germane, and which probably would not be usable as parameters to existing commands and processes anyway. The author's research grapples with this problem, and provides a way to deal with it in as much detail as possible — at each inflection point along a curve.

1.3.2 Encoding Geospatial Data Quality Information

The notion of providing positional metadata for every coordinate location seems to imply that every point in every feature could be sufficiently unique to warrant its own metadata. Given the volume of coordinates in many GIS databases and the rather tiny amount of information that most of them provide, this may seem like an excessive amount of overhead. That, however, can be regarded as an implementation decision, which need not constrain the way in which one thinks about managing and modeling different aspects of geographic space. With this in mind, we shall describe an approach to encoding geospatial data that describes positional certainty independently of the schema used for representing spatial entities. In order to provide a context for this discussion, however, a brief description of logical elements of GIS databases may be useful. Here is a more or less conventional view of geospatial data that has a great many existing implementations in the GIS literature and industry:

Features   = Identifiers + Geometry + Topology + attributes + Metadata

In turn, these terms can be dissected too:
Identifiers   = Names + *Geocode*s + Spatial_Indices
Geometry   = Identifiers + coordinates + Other_Properties
Topology   = Identifiers + *Genus*_Properties + Invariant_Spatial_Relations
coordinates   = Spatial_Metric + Tuples_of_Scalars
attributes   = Identifiers + Data_Items + Metadata
Metadata   = Data_Definition + Data_Quality + Other_Properties

Most GIS databases include most of these elements, but more specialized geospatial applications (desktop mapping packages in particular) may omit or ignore some of them, especially topology and metadata. There seems to be a trend, in addition, in GIS database design to leave out explicit topologic information, replacing it with attribute data, spatial indices or regenerating it as needed, on the fly (ESRI 1996; ORACLE 1995; Jones et al 1994).

Many different implementations of this general approach (or at least some parts of it) exist. The most common approach groups features of a given class together as a *layer* (or *coverage*) which may include topological relations between them but is independent of other feature classes. To relate different classes, layers must be geometrically and topologically combined, a process called *map overlay*. While this thesis is not directly concerned with overlay techniques, it is important to understand their implications for data quality management. Whenever spatial data from different sources are integrated, the lineage of the result becomes heterogeneous. Map overlay works at such a fine-grained scale that many individual features in the resultant layer contain data from two sources (or even more if overlays are cascaded). attributes and metadata for the resultant layer can indicate what sources formed it, and carry over descriptions of their quality, but they cannot easily indicate which portions of particular features came from what source. Therefore, should the positional accuracies of two inputs to an overlay operation differ, the accuracy of the result will vary spatially in uncertain, uncontrolled and undocumented ways. Figure 1.4 illustrates one basis for this pervasive problem.

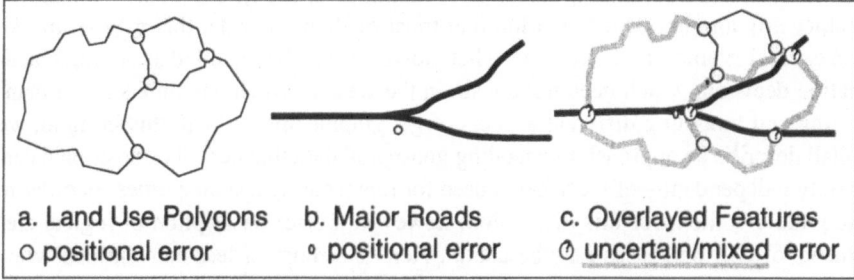

Fig. 1.4: Overlay of data with differing positional accuracy

Some GISs model features as sets of primitive geometric elements ("Primitives") which are stored together or separately, and are linked together to form semantic elements ("features") either via identifiers or topology. For example, the Earth's land masses may be represented as a set of polygons that describe continents, islands and lakes. A motor way can be a set of linked arcs, and a village can be modeled as a set of point features. Even a relatively small feature, such as a family farm, may consist of multiple primitives of several types and include many topological relationships. In general, such "complex features" may consist of groups of points, lines and areas, with or without explicit topology. Obviously, how features are encoded and modeled has strong implications for what operations can be applied to them, and how easy these are to perform. Certainly it is not simple to describe the details of data quality variation where complex features are modeled, just as it is difficult for overlaid data. Alas, there is no optimal data model for geospatial data in general, although there may be nearly optimal ones for restricted sets of data in the context of specific applications. This is one reason why GISs all have different (and usually proprietary) data models, which may be difficult to translate from one system to another.

The problem of documenting heterogeneous data quality is largely unsolved. The U.S. standard for geospatial metadata (FGDC 1994) was a major stride forward in documenting the quality and other aspects of GIS data, as was that country's Spatial Data Transfer Standard (USDOC 1992) before it (the latter was ten years in the making, and is already somewhat obsolete). But both standards deal with data at the dataset level (leaving the scope of datasets up to producers), and make no real provisions for providing finer-grained qualifying data, aside from feature attribute coding conventions. GIS users, of course, are free to embed lower-level metadata in their systems wherever and however they may, normally by creating tables and text associated with specific layers, feature classes and features. This information might be keyed to feature identifiers to link it to geometric objects. The left side of figure 1.5 shows this type of schema, where lower levels of metadata record changes in data quality from the norms for higher levels of abstraction, hence need not be specified for any "normal" object.

There are several problems with this approach, discussed below and in (Dutton 1992). However defined, such metadata records are "auxiliary" data structures, stored separately from descriptions of the geometry and topology of spatial features. Should a set of features be transferred from its host system to another GIS, while metadata records could be provided as well, the receiving system might not be able to use some or all of them, depending on how standardized the metadata records and similar and robust its data modeling capabilities were. Object-oriented systems make this easier to do, but object models must be mapped from one schema to another, not a trivial problem in the general case. Finally, users and producers are burdened with providing (or estimating) data quality information for a complete hierarchy of data elements, not just an entire dataset, unless tools are crafted to determine the metadata and automate their insertion, including formatting them and linking them to associated data entities.

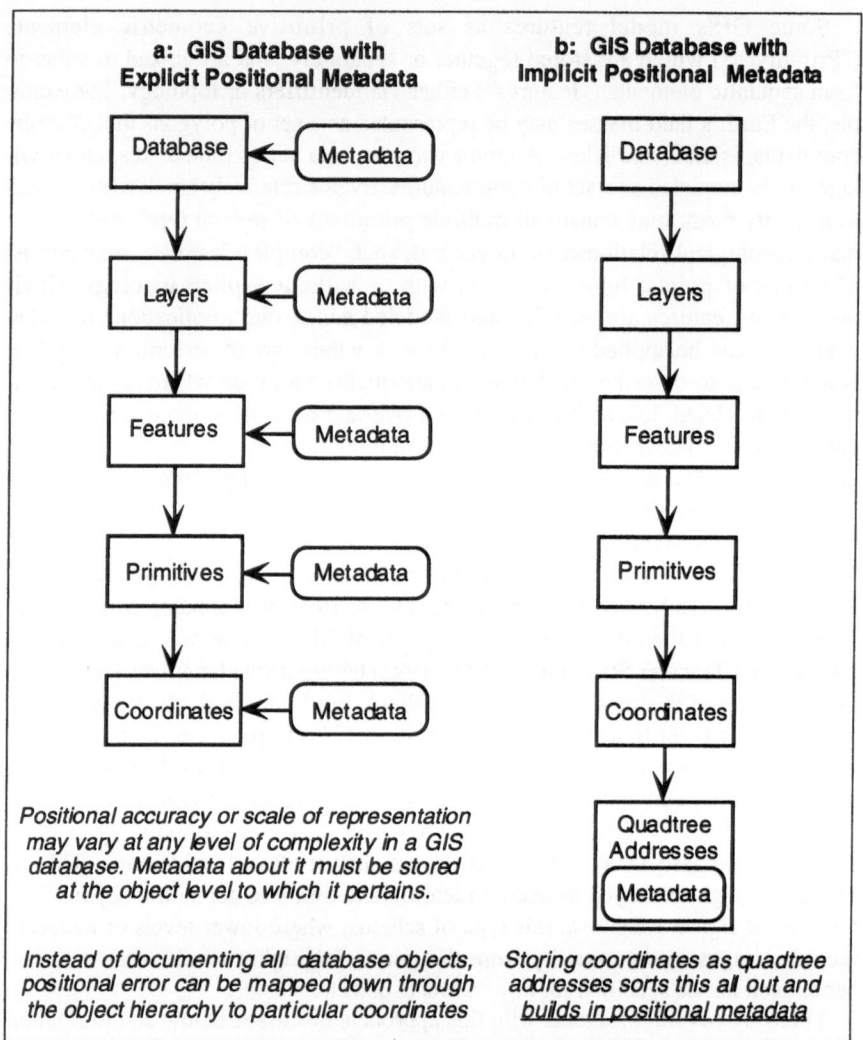

**a: GIS Database with
Explicit Positional Metadata**

**b: GIS Database with
Implicit Positional Metadata**

Database ← Metadata

Database

↓

Layers ← Metadata

Layers

↓

Features ← Metadata

Features

↓

Primitives ← Metadata

Primitives

↓

Coordinates ← Metadata

Coordinates

↓

Quadtree
Addresses

Metadata

*Positional accuracy or scale of representation
may vary at any level of complexity in a GIS
database. Metadata about it must be stored
at the object level to which it pertains.*

*Instead of documenting all database objects,
positional error can be mapped down through
the object hierarchy to particular coordinates*

*Storing coordinates as quadtree
addresses sorts this all out and
builds in positional metadata*

Fig. 1.5: Alternative ways to encode positional metadata (from Dutton 1992)

An alternative way to handle storage of *positional data quality information* (not necessarily all metadata) is illustrated on the right side of figure 1.5, and is described in subsequent chapters. The author believes strongly that this approach provides a much more effective way to embed positional metadata into GIS databases, putting it where it is most useful; at the spatial coordinate level. Properties and consequences of this approach will be explored throughout this document.

1.4 Challenges of Digital Map Generalization

Generalization of digital map data has received an increasing amount of attention in the past several years, not only from the GIS community but from computer scientists as well. This is a multi-faceted problem which will require a multiplicity of approaches, and will keep researchers busy for a number of years to come. Most of the research and solutions in this area are concerned with a single sub-problem, that of *line simplification*, for which a large number of algorithms have been developed. Other work has focused on characterizing point features, aggregating areal features (including administrative units, land use polygons and groups of buildings), text placement, reclassifying thematic values and collapsing feature hierarchies. Besides simplification, other manual techniques (usually called generalization *operators*) that have been identified and automated include *selection*, *smoothing*, *displacement*, *aggregation*, *amalgamation*, *collapse*, *exaggeration*, *enhancement* and *typification* (McMaster and Shea 1992; Brassel and Weibel 1988). These operators are sometimes given different names and definitions in the literature (some may be ignored entirely); professional cartographers are not at all agreed on how to characterize the fundamental activities these terms denote, as a survey by Rieger and Coulson (1993) documented. The most general agreement seems to be that generalization begins with *selection* of data to be displayed, after which other operators come into play.

A special journal issue (Weibel 1995) offers papers from well-known researchers examining a number of aspects of digital map generalization, including line simplification techniques and much more. A recent book from a GISDATA workshop also provides an overview of generalization problems and solutions in the GIS context (Muller et al 1995). Most past work has dealt mainly with the isolated generalization of single features or even portions of features, one or several at a time. But most maps depict more than one feature class or layer, and at a small enough scale a given feature can come into conflict with ones from any other class anywhere in its locality, in arbitrary and complex ways. This has led to a more recent emphasis on *holistic feature generalization* (Ruas and Plazanet 1996), which attempts to assess interactions among features. This is a much different (and more difficult) problem in a vector environment than in a raster-based GIS or image processing system. This is because most map generalization problems result from *competition for map space*, and in vector-based system space is not modeled, phenomena are. Unless features are contiguous, and their topology encoded, extensive analysis may be needed to determine which features are in conflict, especially if they inhabit different classes or layers. In order to handle these kinds of scale-sensitive, combinatorial problems, the trend is for researchers to exploit greater numbers of more complicated data structures, which must all work in concert as they solve specific generalization tasks.

Among the data structuring techniques being investigated to support holistic map generalization are Delaunay triangulations — often constrained by the inclusion of several feature classes at once (DeFloriani and Puppo 1995; Jones et al 1992 and 1994; Ruas 1995). Also being exploited are hierarchical structures for vector data, such as KDB-trees (Robinson 1981), R-trees (Guttman 1984), and many related and derived approaches (Robinson 1981; Cromley 1991; Devogele et al 1996; van Oosterom and Schenkelaars 1995). These techniques are often combined with one another, as well as with other ways to logically integrate and structure multiple versions of map features, such as directed acyclic graphs (Timpf and Frank 1995). The general goal of most of this activity is the creation and exploitation of multi-scale geographic databases, both for cartography and spatial analysis. Jones and Abraham (1986) and van Oosterom (1993) offer overviews of the problems, opportunities and some possible solutions in this increasingly complex area of GIS design.

1.5 Scope of Work

Based on the intuition that generalization of spatial data can be improved by development of multi-scale databases, a specific approach to multi-scale data storage, retrieval and manipulation has been designed, implemented and tested in the course of this research project. A software prototype has been built which, while not a full GIS by any means, provides many of the basic data modeling functions found in one. This testbed has enabled us to assess the essential utility, effectiveness and efficiency of a geoprocessing system based on hierarchical positioning, in the context of performing basic map generalization tasks.

While it is based on a general-purpose, global data model, our project has not attempted to assess its suitability for all GIS and cartographic purposes, only for cartographic line generalization. Nor, as much as we would have liked to, can we claim to have developed more than a partial solution for problems of holistic map generalization. Our fundamental aims are quite specific and relatively modest: *to enrich the content of geospatial data at the coordinate geometry level, so that common operations on data are supported more by the data themselves rather than requiring external metadata, or elaborate post-processing and complex algorithms to cope with missing information.*

Still, there are some senses in which our approach is more universal than existing vector and raster data architectures: it is global (planetary) in scope, and represents features at a hierarchy of resolution (scales), properties which make it applicable to a broad range of geospatial applications, not just world-wide ones, and not simply for cartographic purposes. And while the new generalization methods developed by this effort seem to work quite well, perhaps the basic contribution to map generalization is more diagnostic than prescriptive. Using the a hierarchical framework may make it easier to detect and negotiate conflicts for map space, but additional hard work will be needed to translate these insights into working systems that yield a long sought-after solution: generating acceptable map products across a range of scales from a single detailed geospatial database.

The potential utility of the approach to location modeling described in this thesis extends beyond map generalization, to indexing of point, vector and raster terrestrial and astronomical data, modeling global environmental phenomena and processes, and multi-scale terrain modeling. All of these areas have been addressed by researchers exploring hierarchical polyhedral planetary representations, some derived from the author's work, others closely related to it. The scope and context of these investigations are summarized in the next chapter, following a descriptive overview of our specific approach to modelling geographic location. Differences in purposes and backgrounds of cited researchers cause some inconsistencies in the organization and nomenclature of their models, which are all basically quite similar, varying only in a few basic respects.

Chapter three descends into greater detail to describe salient geometric, topological and numerical properties of the model. Details concerning notation and coordinate conversion are presented (pseudocode for the latter is provided in appendix A). Also dealt with is *spatial indexing*; a derived hierarchical construct for this purpose is introduced and discussed in this context, with some implementation details provided in chapter 4. Use of our model to index global point data is summarized in appendix E. Chapter 3 concludes by illustrating how hierarchical methods can both ascertain and verify important aspects of positional data quality for cartographic data.

Chapter 4 narrows the focus to map generalization, describing a feature-oriented data model and some hierarchical algorithms designed for characterizing and filtering cartographic detail. It starts with an overview of digital map generalization, providing a context for our own approach to line simplification. Also discussed are assessing and conditioning line data and approaches to identifying conflicts among map features due to scale change. Details of an operational data model are presented, followed by those of a line simplification algorithm for hierarchical coordinates and the control parameters it uses. The chapter closes with an explanation of the generation and handling of point-specific attributes (metadata), showing how they are used to guide point selection for line generalization.

Chapter 5 describes how generalization methods presented in chapter 4 were implemented, and along with appendices C and D, presents results of generalization experiments using two shoreline datasets. The software and hardware environment used for development and testing is described, as well as the methods used for data handling and presentation. Results of processing each of the test files are discussed and illustrated, followed by an evaluation of strengths and weakness of our methods in relation to the dominant line simplification paradigm.

In chapter 6, we summarize what our empirical studies seem to say about a major tenet of cartographic generalization, the notion of *characteristic points*, looking at it from several perspectives, including scale imperatives, point weeding, local and global generalization strategies and evaluation of results from the algorithms compared in chapter 5. We then propose a counter-intuitive reconstruction of this notion, along with a way to implement it. The chapter concludes with a research agenda for making the hierarchical coordinate approach a more robust basis for multi-scale database construction and automated map generalization.

1.6 Chapter 1 Summary

This chapter has attempted to describe types difficulties that can result from poor or undocumented positional data quality. These need to be overcome in order to properly integrate, update, analyze and display geospatial data in a GIS environment. While many aspects of data quality obviously affect these activities, the research reported here is focused on improving the descriptive power of locational notation. In particular, the assertion is made that reliance on spatial coordinate tuples, such as latitude and longitude, constrain the ability of software to manage geodata, and that it is possible to minimize this shortcoming by using a hierarchical approach. The particular application focus of the research, cartographic generalization, has been chosen to demonstrate how map scale and spatial resolution can be linked in a data model that documents the precision of feature locations and permits them to be accessed across a range of spatial resolutions and map scales.

The remainder of this first chapter focused on how digitization of and coordinate notation for spatial geodata hampers its use. First, several types constraints arising from the process of discretization were described; limitations of scale of capture and the inherent numerical precision of computer hardware and software both are shown as contributing to lack of certainty about location. The next section discussed measuring and documenting positional data quality. After presenting an overview of typical elements of GIS datasets, a set of reasons were described that contribute to poor documentation of spatial accuracy in geodata, and why this is inevitable given how such data are conceived of and structured. These problems were put into focus by considering the difficulties in generalizing digital map data, in which locational uncertainty may vary within features in ways that data structures are unable to document. This was further addressed in a brief review of current map generalization research directions, in which one common task is to add contextual information to map data, by enriching temporary or archival data structures; this focus on data enrichment is maintained throughout the thesis. The chapter ended with a plan of the thesis, indicating what the project has sought to accomplish, describing topics covered, problems addressed and solutions presented in each subsequent chapter.

2 Historical and Conceptual Background

Everything is where it is having moved there.
— William Warntz

For hundreds of years, *locations* on the Earth have been described using lines of latitude and longitude. In advancing their fields, mapmakers, mariners, geographers, surveyors and many others have relied upon and refined this geodetic planetary grid to a high level of precision. With the marriage of aerospace and computer technology, the determination of location has become an advanced art, and serves human societies in many and diverse ways. Although many people who develop and use geographic information systems are quite unaware of it, reliance on latitude and longitude positions — and projections of them — has hampered the ability of their software to process and map locational data. Something is missing, something that would make it easier to deal with problems of scale, accuracy and data quality.

While a latitude and a longitude can pinpoint where something is on Earth, it can't say how big that thing is or how accurate the estimate of its location may be. Although the precision of such measures may be known to their makers and users, computer software may lose track of positional data accuracy, since all coordinates of a given type tend to be represented to uniform hardware precision (usually 16, 32 or 64 bits). As was discussed in chapter 1, one can always document the accuracy of coordinates and the sizes of things they locate separately and apply this information when processing the coordinates. But a typical GIS does not make this easy to do, and certainly not automatic.

But most GIS databases do not even record locations in latitude and longitude; planar projections are almost always employed. The reasons for this are various, but an important one is that primary collectors of geodata — military organizations and other government agencies — often prefer to survey and map in plane coordinate systems having a standard projection, such as the Universal Transverse Mercator (UTM) and a standard vertical datum, such as WGS84. Most nations have an official grid system (or several) for surveying and cadastral purposes. The grid may contain a set of zones, each a local coordinate origin. In principle, any grid location can be *georeferenced* back to the sphere, in order to relate it to geospatial data not encoded in official grid units.

Georeferencing a planimetric location requires a number of pieces of information: (a) the zone in which the location falls; (b) the latitude and longitude of the origin of the zone; (c) the projection type; (d) the parameters used to create the projection; and (e) the datum (ellipsoid model) used. Most national grids use projections having inverses, enabling reconstruction of latitude and longitude from grid locations, with a (presumably) documented, relatively uniform accuracy. Deprojection can then be automated, perhaps consuming a noticeable amount of time, but otherwise relatively transparently. As Chapter 1 discussed, coordinate transformations using floating point numbers are rarely lossless, and the amount of error or noise that projection and deprojection algorithms add to coordinates may need to be accounted for.

GISs also tend to use planar coordinate systems because many of their processing functions assume that coordinates and distances are Cartesian rather than spherical. Many such functions require distance computations (and comparisons between them) in order to produce valid output, and all display functions require mapping to the plane. The general consensus seems to be that a well-defined projection is preferable to geographic coordinates, which only come into play when integrating new data — for example, Global Positioning System (GPS) measurements — or when changing projections for data integration, display or export. In the early days of GIS, there were few sources of digital map data, which were generally digitized from maps that may or may not have had well-documented coordinate systems. Consequently, spatial data were often not reusable by other applications, even for different projects in the same organization. This situation has much improved over the last decade, but still today, even with abundant spatial metadata resources to document datasets and software that can reconcile coordinate systems, many lapses in data quality go undetected until far too late in the cycle of data reuse, if ever.

2.1   A Typology of Spatial Data Models

Each GIS has its own *data modeling* philosophy that is likely to be only in partial agreement with those of others. A system's architecture limits what types of phenomena are representable by specifying the abstraction mechanisms used to encode them. Within these limits, considerable freedom may be provided to database architects (users who design database schemata for GIS projects) in the ways that real-world entities and phenomena may be described and quantified. Decisions must be made, for example, whether to put hydrographic features on one layer, or partition them into layers for lakes, streams and other water bodies. In a raster-based system, if a fixed cell size is required, the extent of cells may need to be decided upon, requiring negotiations among various themes having differing spatial resolution. In designing vector GIS databases, it may be optional to model containment relations (e.g., point-in-*polygon* or polygon-in-polygon), and whether this is done will have implications for processing later on. Lastly, object-oriented databases provide a great deal of freedom (perhaps too much) in specifying data models. Absent standards and guidelines, one *object model* for a given set of phenomena may look very different from another. Such models depend on how source data is encoded, the applications for which the database is designed, and the ways in which database designers conceptualize phenomena.

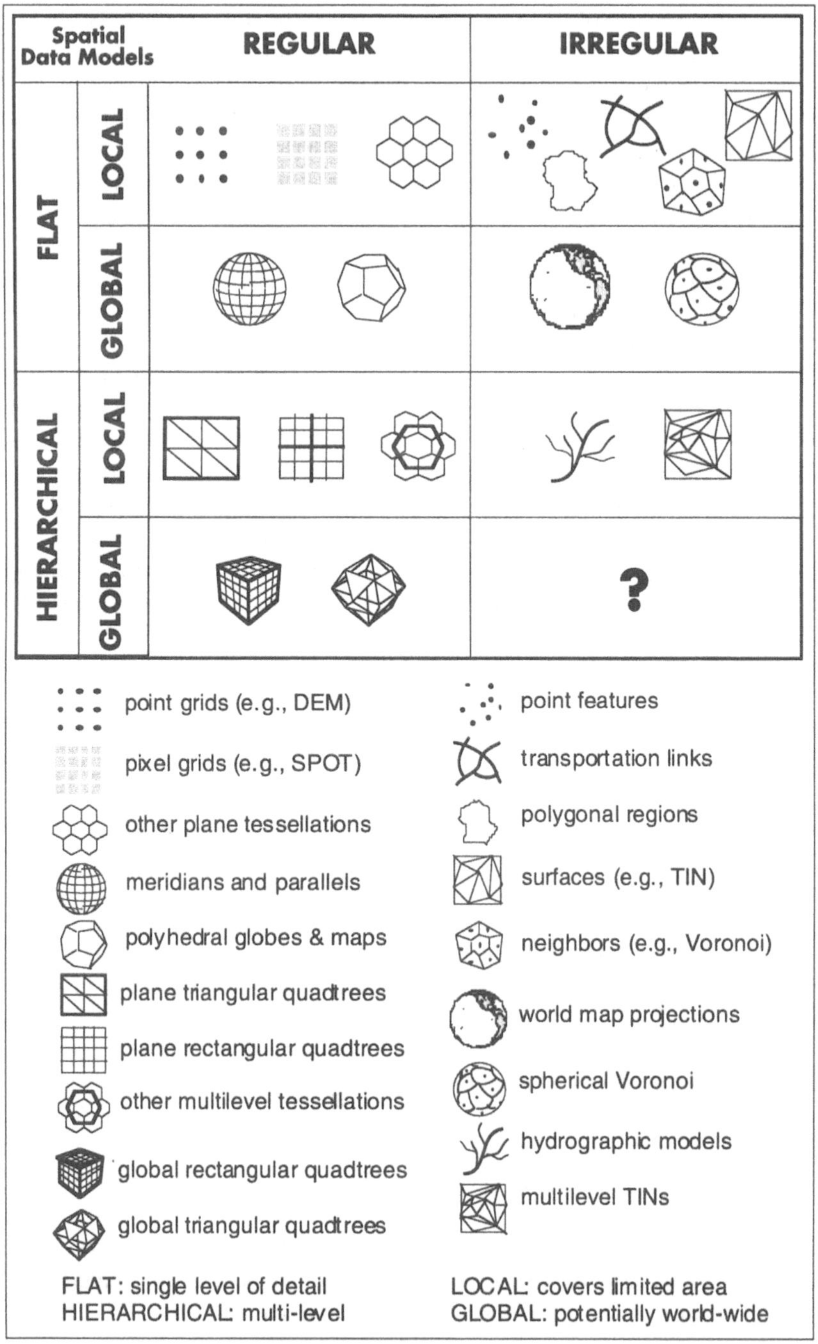

Spatial Data Models	REGULAR	IRREGULAR
FLAT — LOCAL	point grids, pixel grids, other plane tessellations	point features, transportation links, polygonal regions, surfaces
FLAT — GLOBAL	meridians and parallels, polyhedral globes & maps	world map projections, spherical Voronoi
HIERARCHICAL — LOCAL	plane triangular quadtrees, plane rectangular quadtrees, other multilevel tessellations	hydrographic models, multilevel TINs
HIERARCHICAL — GLOBAL	global rectangular quadtrees, global triangular quadtrees	**?**

point grids (e.g., DEM)

pixel grids (e.g., SPOT)

other plane tessellations

meridians and parallels

polyhedral globes & maps

plane triangular quadtrees

plane rectangular quadtrees

other multilevel tessellations

global rectangular quadtrees

global triangular quadtrees

point features

transportation links

polygonal regions

surfaces (e.g., TIN)

neighbors (e.g., Voronoi)

world map projections

spherical Voronoi

hydrographic models

multilevel TINs

FLAT: single level of detail
HIERARCHICAL: multi-level

LOCAL: covers limited area
GLOBAL: potentially world-wide

Fig. 2.1. An iconic typology of spatial data models

It is not the purpose of this thesis to dissect specific data models used in GIS, or argue for a particular approach, although aspects of data modeling do figure in the following chapters. In fact, a specific data model for representing cartographic features is proposed in Chapter 4, one that attempts to combine a feature-based approach with the vector-topological model that underlies most non-raster GIS, all under-girded by hierarchically-encoded coordinates. It is one way to enrich map data to make map generalization problems easier to identify, formalize and hopefully solve. But it is only one way, and it is only a start.

Before giving details of our approach in section 2.2, it may help to describe a context within which it can be interpreted and compared to related approaches. We first present the context in the form of a classification of spatial data models employing three dichotomous categories. In this typology — similar to ones by Peuquet (1984) and Kainz (1987) — different approaches to structuring spatial data are distinguished according to whether they are:

• *Regular* or *Irregular* — Divide space or phenomena uniformly or arbitrarily
• *Local* or *Global* — Intended to a cover limited region or the entire Earth
• *Flat* or *Hierarchical* — Describe one level of detail or multiple levels

Figure 2.1 summarizes this typology iconically, displaying and describing typical representations of seven of its eight classes. Some of the classes described are obscure or specialized, others more generic. Nor are they necessarily mutually exclusive; GIS databases may include several two-dimensional datatypes and even be able to relate them. Some examples predate geoprocessing, and serve to remind us that not all data models are digital. Digital or analog, current or historical, it is still instructive to see how these inventions relate. For the most part, vector-topological GISs model spatial data as "irregular/local/flat" while raster GISs and image processing systems operate on "regular/local/flat" data. In terms of technologies, the "regular/irregular" dichotomy tends to predict whether a system models *space* or objects primarily (e.g., raster or vector). Some approaches manage hierarchies of objects, others objects or divisions of space at a *hierarchy* of resolution, or at least incorporate hierarchical *spatial indexing* mechanisms even though objects themselves are not encoded in a hierarchical fashion; see discussion of (Oracle 1995) in section 3.2.

To our knowledge, *no general-purpose, commercial, vector-based GIS uses hierarchical coordinates to encode locations*, allowing retrieval of map features at multiple resolutions, as proposed in this thesis. Thus, the only class in figure 2.1 that is not occupied — irregular/ global/hierarchical models — could exist, but seems not to have been implemented in a GIS to date. An example of this class would be a hierarchical spherical Voronoi *tessellation*. This would be a logical extension of work by Lukatella (1989) and Watson (1988) on methods to compute spherical *Voronoi diagrams*; however, over many levels of detail, management of Voronoi geometry and topology could become cumbersome.

But the eighth category could certainly be populated using the concepts we will be describing. The approach presented here a bit unusual — causing it to cross over categories in the above typology — in that it is a *regular* division of space applied to handling *irregular* spatial objects; i.e., it uses a *"space-primary"* data model to encode an *"object-primary"* world. It is also notable for the link it forges between *resolution* (a digital data parameter) and *scale* (an analog map parameter). Tobler (1988) is probably responsible for first making this connection explicit, but its reduction to GIS practice is long overdue.

2.2 A Global, Regular, Hierarchical Spatial Data Model

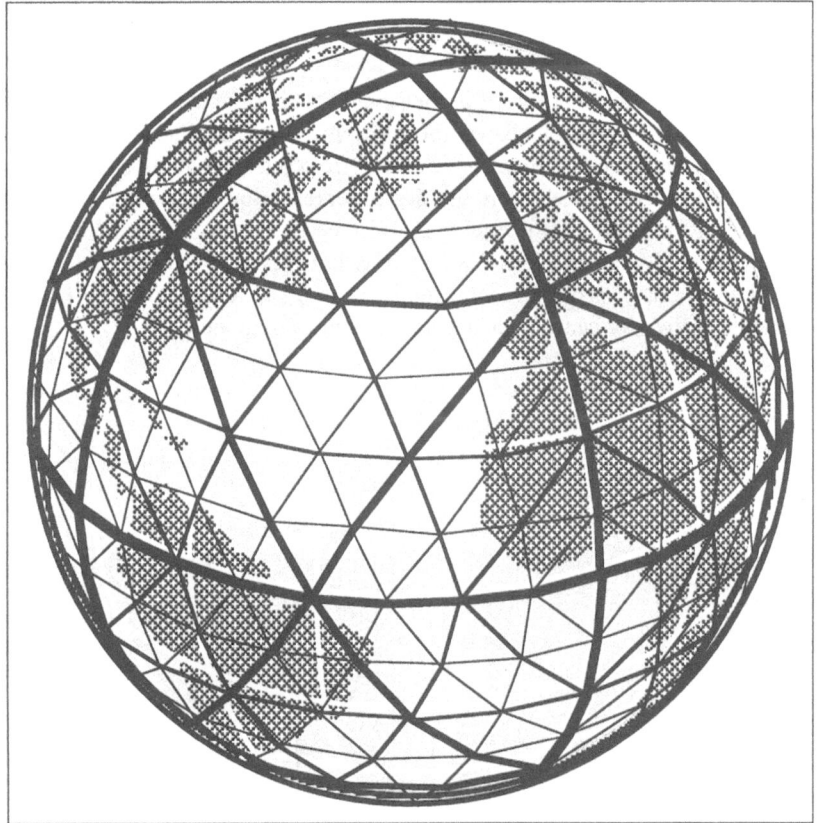

Fig. 2.2. Quaternary Triangular Mesh to Three Levels of Detail

In place of latitudes, longitudes and metadata, we propose a notation for location that documents its own precision, one which can model spatial and scale relationships as well as denoting positions. Data in this form can be derived from coordinates plus scale or accuracy information, and is thus compatible with existing digital georeferenced data. Once so encoded, positions can be retrieved at a wide range of precision, i.e. at specific scales or resolutions. Locations neighboring a given one can be directly identified at any resolution. Every point on the planet has its own unique hierarchical geocode, accurate to the extent its location can or needs to be measured. Our notation uses integer *identifiers* rather than floating-point coordinates, and at first sight do not appear to be coordinates at all. But they directly denote position, and do so at multiple, nested resolutions. This is why we refer to it as a *hierarchical coordinate system*, the form and orientation of which is described by figures 2.2 and 2.3

These properties are achieved through the use of a geometric computational model that, starting with an *octahedron* we imagine to be embedded in the Earth, subdivides its eight *faces* into four each, proceeding to recursively subdivide each of the 32 resulting triangular facets (also called *quadrants*) into four children. The basic geometry and topology of this hierarchical model are illustrated in figure 2.3. While all facets can be subdivided in the same way, in practice this is not done everywhere, but only where one has point positions to encode. Each such location is identified by a sequence of *quaternary digits* (from 0 to 3), having as many places as the number of times facets were divided in representing that point. We call each of these places a *level*, as it is equivalent to a certain *level of spatial resolution*. Our implementation allows for 29 levels of detail (in a 64-bit word), corresponding to about 2 cm of ground resolution, or to maps at a scale of 1:50 or larger. This should suffice for nearly all cartographic and analytic purposes.

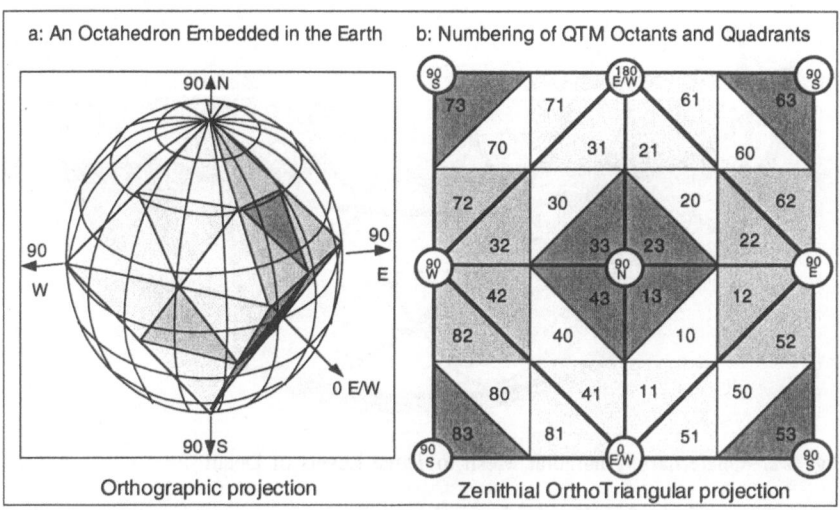

Fig. 2.3. Form and orientation of the quaternary triangular mesh

Detailed images of this structure are difficult to render properly, due to the great number of subdivisions that quickly accumulate. But many of its hierarchical properties are readily summarized numerically; table 2.1 presents the major ones.

QTM LVL	ID BITS	FACETS/ OCTANT	FACET AREA *	UNIT	LINEAR DIVISIONS	LINEAR EXTENT	UNIT	SCALE DENOM.	"STD" SCALES
1	6	4	16,120,936	kmsq	2	5,000	km	1.00E+10	
2	8	16	4,030,234	kmsq	4	2,500	km	5.00E+09	
3	10	64	1,007,559	kmsq	8	1,250	km	2.50E+09	
4	12	256	251,890	kmsq	16	625	km	1.25E+09	
5	14	1,024	62,972	kmsq	32	313	km	6.25E+08	
6	16	4,096	15,743	kmsq	64	156	km	3.13E+08	
7	18	16,384	3,936	kmsq	128	78	km	1.56E+08	
8	20	65,536	984	kmsq	256	39	km	7.81E+07	1:100M
9	22	262,144	246	kmsq	512	20	km	3.91E+07	1:50M
10	24	1,048,576	61	kmsq	1,024	10	km	1.95E+07	1:20M
11	26	4,194,304	15	kmsq	2,048	5	km	9.77E+06	1:10M
12	28	16,777,216	4	kmsq	4,096	2	km	4.88E+06	1:5M
13	30	67,108,864	960,883	msq	8,192	1	km	2.44E+06	1:2M
14	32	2.6844E+08	240,221	msq	16,384	610	m	1.22E+06	1:1M
15	34	1.0737E+09	60,055	msq	32,768	305	m	6.10E+05	1:500K
16	36	4.2950E+09	15,014	msq	65,536	153	m	3.05E+05	1:250K
17	38	1.7180E+10	3,753	msq	131,072	76	m	1.53E+05	1:100K
18	40	6.8719E+10	938	msq	262,144	38	m	7.63E+04	1:63K
19	42	2.7488E+11	235	msq	524,288	19	m	3.81E+04	1:50K
20	44	1.0995E+12	59	msq	1,048,576	10	m	1.91E+04	1:25K
21	46	4.3980E+12	15	msq	2,097,152	5	m	9.54E+03	1:10K
22	48	1.7592E+13	4	msq	4,194,304	2	m	4.77E+03	1:5K
23	50	7.0369E+13	1	msq	8,388,608	1	m	2.38E+03	1:2.5K
24	52	2.8147E+14	2,291	cmsq	16,777,216	60	cm	1.19E+03	1:1K
25	54	1.1259E+15	573	cmsq	33,554,432	30	cm	5.96E+02	1:500
26	56	4.5036E+15	143	cmsq	67,108,864	15	cm	2.98E+02	1:300
27	58	1.8014E+16	36	cmsq	134,217,728	7	cm	1.49E+02	1:150
28	60	7.2058E+16	9	cmsq	268,435,456	4	cm	7.45E+01	1:75
29	62	2.8823E+17	2	cmsq	536,870,912	2	cm	3.73E+01	1:40
30	64	1.1529E+18	1	cmsq	1,073,741,824	1	cm	1.86E+01	1:20

Table 2.1. Quaternary triangular mesh statistics for 30 levels of geometric detail; note that *facet area* and *linerar divisions* are both averages for the globe, as are scale factors

In table 2.1, *QTM Level* (col 1) is the (recursive) depth of subdivision; *ID Bits* (col. 2) gives the size of facet Identifiers in bits. The third column refers to the number of subdivisions in an **octant** (1/8 of the earth); the *Linear Divisions* column likewise specifies how many times an octant's **edges** are subdivided at each level. Dividing that number into one quarter of the earth's circumference (10,000 km) yields the average length of a facet edge, labelled *Linear Extent*. Since these lengths do vary (by up to 35%, as section 3.1.1 shows) they are presented as round numbers and scaled to be easier to comprehend. The last two columns present a rough correspondence between spatial resolution and map scale. This general relation is based on an (arguable) assumption that the positions of points on a printed map are accurate to within 0.4 mm (Tobler 1988). Such accuracy exceeds U.S. national map accuracy standards (see section 1.3), but is still coarser than some line weights on topographic maps. The *scale denominator* is computed by dividing *Linear Extent* (in mm) by 0.4 mm. Simplifying further, the quotient is rounded to a whole number in the *Standard Scales* column.

As space is recursively subdivided into four nearly similar units, we have, computationally speaking, a *quadtree*, or quaternary division of space. More precisely, they form a *forest* of eight *region quadtrees*, one for each octahedral face, or *octant*. Region quadtrees are but one of numerous variants of hierarchical space partitioning systems, but are perhaps the most widely used of them. In our approach, however, as *vertices* (point locations) are encoded (as opposed to line elements or polygons), we could also call the data structure a *point quadtree*. In Samet's semi-official terminology, we could call our storage mechanism a *PR Quadtree* (P for point, R for region), as each leaf *node* is a region that is occupied by one and only one vertex (Samet 1990: 92).[1] Further-more, rather than a rectangular division of space (which quadtrees almost always employ), we use a three-axis system; it is *triangles* (starting with a set of eight) that subdivide into four each. The edges of these triangles are aligned in a finite three-way *mesh* that extends around the planet and densifies where precision goes up. Hence the model's name, *quaternary triangular mesh*, or QTM[2] for short.

2.2.1   Local and Global Properties

We orient the octahedron to the Earth by aligning its vertices with cardinal points — to the poles and at longitudes 0, 90, 180 and -90 degrees along the equator — and number the octants from 1 to 8, as defined by table 2.2. The triangular subdivisions of octants are identified by adding a quaternary digit (0, 1, 2 or 3) to a series of them built up in the course of fixing a location. Upon reaching a point's limit of precision the process stops, and the sequence of digits becomes the *QTM address* (a quadtree leaf node which we denote as *QTM ID*) of that point. Address digits are assigned to quadrants according to a method that always numbers the central one 0, but permutes the other three identifiers at each level. This method results in grouping six 1's, 2's or 3's around every grid node, reflected in the last digit of facets' QTM IDs.[3]

---

[1] Strictly speaking, the single-node-occupancy restriction is only enforced on a per-object basis. In our implementation, it is entirely possible for different objects to share leaf quadtree codes; during map generalization, this happens all the time.

[2] The pun on UTM is intentional...

[3] In the QTM tessellation, every node in the mesh has six triangular facets incident to it, except for the six vertices of the initial octahedron, each of which has four neighbors. This can be taken into account when finding neighbors, for example, but it is not often required as five of the six fourfold vertices are in ocean areas and the

An octahedron has three axes that connect poles, and three orthogonal "equators." Our way of coding its facets yields 4^{29} (2.88230...E17) child facets. Every facet is defined by three mesh nodes at which (with six exceptions) six facets meet. When a node lies on the edge of an octant, three of its incident facets lie in one octant's quadtree; the other three are stored in an adjoining one. A similar relation holds true for any node, not just those along the edges of octants; facets incident at a node are cousins, not siblings, and relating objects that span them can be awkward in quadtrees. Because they relate quadrants in different trees and sub-trees, nodes play a critical role in the QTM framework, and are important enough to warrant the special name *Attractor*.[4] For many purposes, attractors are as important for retrieving locational data as QTM addresses are for encoding them. Properties and uses of attractors will be examined in detail in following chapters.

Octant ID	Minimum Longitude	Maximum Longitude	Minimum Latitude	Maximum Latitude
1	0	< 90	> 0	90
2	90	<180	> 0	90
3	180	<270	> 0	90
4	270	<360	> 0	90
5	0	< 90	-90	0
6	90	<180	-90	0
7	180	<270	-90	0
8	270	<360	-90	0

Table 2.2. Orientation of the QTM basis octahedron to the earth

In describing QTM, we have so far stressed its global properties, as these are what qualify it as a (hierarchical) coordinate system suitable for encoding geographic data. Based on this overview, some readers might assume that QTM is only intended to handle geospatial data which spans the globe, such as world climatological, bioregional or demographic statistics. In the large, QTM can certainly support such world-wide applications, but only because it also works in the small, and throughout a range of scales. In this project we have deliberately focused on the small end of this continuum, but not simply to avoid grappling with global datasets, which would tax our processing platforms without adding much explana-

sixth is in Antarctica, at the South Pole.

[4] The term is borrowed from mathematical chaos theory, which also provides the concept of *strange attractorr*. It is normally taken as a location in k-space (often in the complex plane) around which a function may oscillate or orbit, although it may never actually occupy the attractor's location. In the QTM model, every node "attracts" the facets that touch it at a given level, hence all locations in that set of facets "belong" to that attractor, which may be used to spatially relate them.

tory value; we have done this to explore essential details of hierarchical data handling that are critical to understand, regardless of the scope of one's spatial data and their application.

2.3   Prior, Related and Derived Work

The quaternary triangular mesh geometric data model is derived from and marries a number of concepts, principally polyhedra and quadtrees. While the latter are products of the digital era, going back about 25 years, the former are very old, and even their application to cartography has a long history. The following sections point out some similarities between QTM and earlier polyhedral maps, and trace the use of such constructions as models for storing digital environmental data. The discussion concludes with descriptions of data models that have been derived from QTM and related sources.

2.3.1   Polyhedral Map Projections

Mapmakers and others have modeled the earth as a ***polyhedron*** for many years, going back to at least the time of the German artist Albrecht Dürer (1471-1528), whose drawings of polyhedral globes appear to be the first instance of thinking about mapping the planet in this way. Many more of them subsequently were invented.

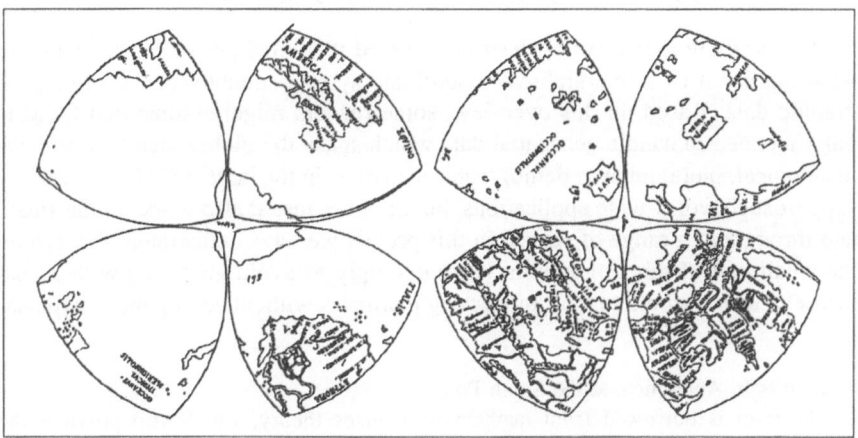

Fig. 2.4. Da Vinci's *mappemonde* from c. 1514 (Nordenskiöld 1889: 77)

A rather remarkable historical world map found in Leonardo Da Vinci's papers is reproduced in figure 2.4. The source of the illustration is A.E. Nordenskiöld's (1889) atlas of early cartography. About this map, that author remarks:

A good representation of the geographical ideas prevailing in the period immediately preceding Magellan's circumnavigation of the earth, is found in a collection of Leonardo Da Vinci ... From this circumstance a certain interest is attached to this insignificant sketch, which is in no wise distinguished by such accuracy and mastery in drawing, as might be expected from a map attributed to the great artist among whose papers it was found. It is, however, worthy of attention from a cartographical point of view, not merely on account of the remarkable projection, never before employed, but also because it is one of the first maps on which a south-polar continent is laid down. It is likewise, if not the first, at least one of the first mappemondes with the name America. (Nordenskiöld 1889:76-77)

Whoever made this sketch was clearly copying a globe, yet the octahedral symmetry in double polar aspect is still startling. The east-west orientation places meridian zero not in the Atlantic, as might be expected (many maps from the period show a prime meridian at about 10° W), but roughly through Lisbon, indicating the globe might have been of Portuguese origin. Following this unheralded sketch, no octahedral projections seem to have been published for more than 300 years. However, in the late 19th and early 20th centuries, various cartographers, such as Cahill (Fisher and Miller 1944) rediscovered this idea, projecting the land masses of the Earth to various polyhedra, then unfolding their facets into flat, interrupted maps. Figure 2.5 shows such a projection, this one on a cube, and fig. 2.6 displays Clarke's octahedral Modified Collignon projection (Clarke and Mulcahy 1995).

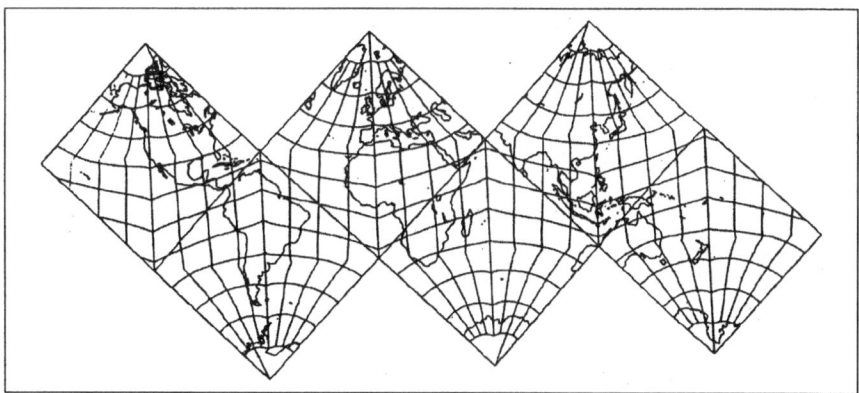

Fig. 2.5.   Hexahedral world projection, vertices at poles (Snyder 1992: 17)

The best known polyhedral projection today is probably R. Buckminster Fuller's *Dymaxion map*, dating from the early 1940's (Unknown, 1943). As described in that Life Magazine article, the Dymaxion map was first based on a cubeoctahedron. A few years later Fuller recast it as an ***icosahedron***, oriented to the earth in a way that placed all 12 of its vertices in the oceans and minimized the

division of land areas between its 20 facets. Fuller devised a projection method —
only recently well-enough understood to implement digital algorithms for it (Gray,
1994; Gray 1995; also see Snyder, 1992) — that has remarkably little distortion.
Fuller's motivation was twofold: to create a map projection having as conformal
and equal-area as possible, and to develop an inexpensive, easily fabricated and
transported globe that could be used for displaying world-wide thematic data. While
die-cut cardboard versions of the Dymaxian map are still available, it has generally
been the fate of Fuller's and most other polyhedral projections to become curiosi-
ties and playthings for aficionados.

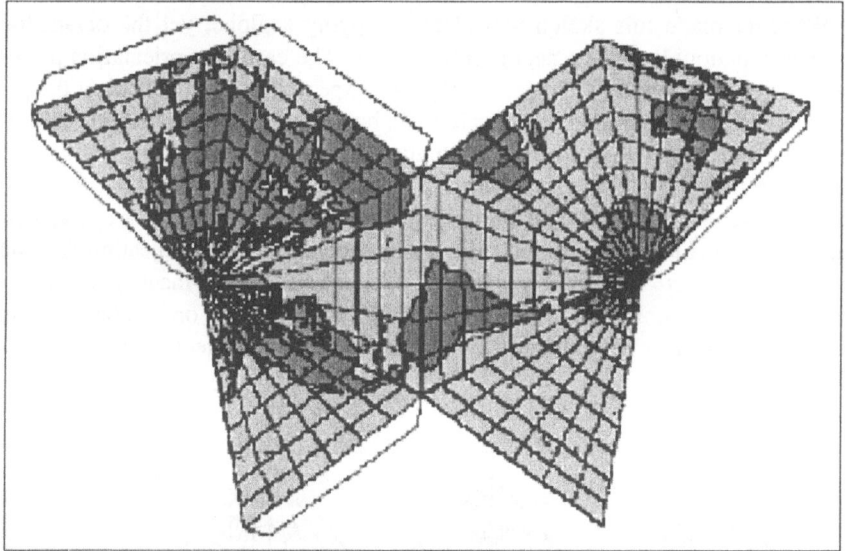

Fig. 2.6. Clarke's Modified Collignon ("Butterfly") Projection
© copyright Keith C. Clarke 1994)

In the digital realm, almost all applications of global tessellations involve the
use of map projections, some examples of which are given above. QTM data processing (including our own and related work described in section 2.3.4) is no ex-
ception to this. We make use of a map projection (Dutton 1991) when transforming QTM addresses from and to latitudes and longitudes, called the Zenithial
OrthoTriangular projection (ZOT), as it is an azimuthal projection that creates
uniform right triangles out of QTM facets (which are not uniform on the sphere).
The projection can be centered on any octahedral vertex (we use a North polar
aspect), and maps an octahedron to a square in which the opposite (South) pole
occupies the four corners and the equator connects the centers of sides.

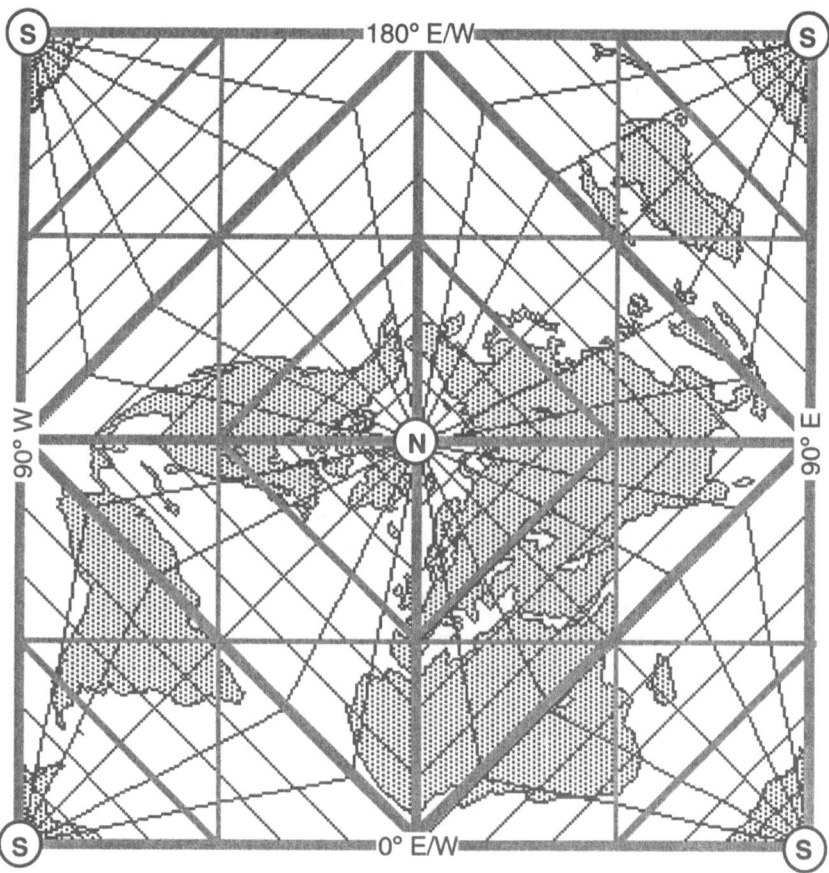

Fig. 2.7. Zenithial OrthoTriangular World Projection, with Geographic and QTM Graticules

All parallels and meridians are straight lines in ZOT, and all QTM facets are nested, right isosceles triangles. This comes from using *Manhattan metric* (rather than Euclidean distance), in which distances between points are absolute sums of X and Y displacements. Figure 2.7 illustrates the projection for the globe; grey lines show level 1 QTM facets (which subtend 45°) in relation to a 15° geographic *graticule* (in black). It is obvious that ZOT produces many distortions; it is neither conformal nor equal-area. Other than being doubly-periodic (it can be tiled to fill a plane), its cartographic advantages are minimal, and is only used because it makes QTM addressing so simple and fast. Refer to appendix A for details on ZOT projection arithmetic.

2.3.2   Global Environmental Monitoring

Until quite recently, polyhedral maps received little attention in professional cartographic circles, and even less in the GIS arena. However, a small amount of more serious interest has been evident in the literature of ecology and geosciences, in which the need for global sampling of environmental data periodically arises. The earliest digital implementation using polyhedra seems to be the one reported in Wickman et al (1974). Requiring a sampling regime for global geochemical data that was as uniform as possible, and recognizing that five platonic polyhedra are the only figures that are completely uniform tessellations of a sphere, Wickman compromised, opting for a triangulation of a *dodecahedron*. This figure consists of 12 regular pentagons, and can be divided into 60 identical 72-60-60-degree triangles by connecting the center of each pentagon to its vertices. Each of these triangles was then subdivided into four child triangles by connecting its edge midpoints by chords. On a sphere, such a subdivision does not lead to four triangles having equal area. The central one is always largest, although the relative differential in area shrinks rapidly at higher levels of subdivision. Wickman's group preferred true equal area partitions, achieved by inserting "correction points" near the center of each chord, turning each edge from one into two great circle arcs, and positioning these points to make the resulting facets equal in spherical area. A numbering scheme was then devised that encoded the tessellation hierarchically with the digits 1-4. The system was coded in Fortran IV on an IBM 360/75 computer.

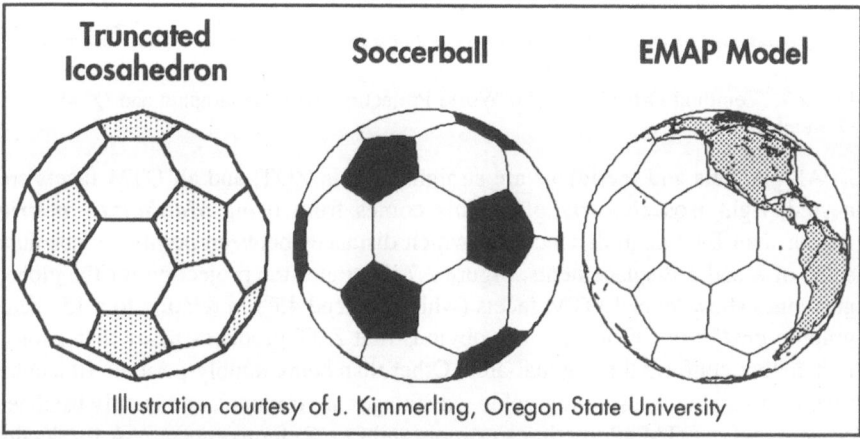

Fig. 2.8. EMAP Polyhedral Global Sampling Model

A more recent example in the same spirit is described in White et al (1992), summarizing a hierarchical data model developed for the U.S. Environmental Protection Agency's EMAP (Environmental Monitoring and Assessment Program) effort. Their goal was to devise an environmental sampling regime that could work anywhere in the world (but would only be initially implemented for the U.S. and its territories), which would have equal area hierarchical sampling units. Statistical considerations (minimizing distance variance between sample points) pointed toward using a basis polyhedron with a large number of facets, and the one selected (a truncated icosahedron, familiar as a *soccer ball*, and also as the **Fullerene** carbon molecule) has 32 of them, 20 hexagons and 12 pentagons. Figure 2.8 shows this shape, its soccer ball variant, and how it is oriented for use in EMAP. This particular orientation allowed the 48 conterminous U.S. states to fit in one hexagonal facet (Alaska occupies another hexagon and Hawaii a third one); but inevitably, this strategy dissected other nations and territories in inconvenient ways. Each of the 32 facets in this model is treated as flat, and all locations within it projected from geographic coordinates to planar ones using a Lambert conic equal area projection that varies distances between sampling points by less than two percent (White et al 1992).

Within any primary EMAP hexagonal or pentagonal plate, various tessellations provide equal-area sampling units across a range of scales. Triangular, rhombic and hexagonal grids may all be utilized, depending on the type of resources and stressors to be monitored and the desired sampling density. The exact shape and hierarchy matter less than does the equality of sampling units, and may be chosen to be appropriate to the monitoring task. Relating data from neighboring polyhedral facets is somewhat awkward, but the model was devised to minimize the necessity of doing this.

2.3.3 Global Terrain Modeling

The research reported here stems from a project in the early 1980's to create an efficient data structure for global relief modeling. Initially, this was a data compression technique designed for (local rather than global) digital elevation data in gridded form.. A lossy algorithm was developed that would compress a digital elevation model (*DEM*) from 16 or 32 bit per value to 2.67 bits. The scheme, called *DEPTH* (for Delta-Encoded Polynomial Terrain Hierarchy) was based on the assumption that grid samplings of continuous *surfaces* such as terrain have vertical accuracies that are constrained by their horizontal sampling resolution; the larger an area covered by a grid cell, the less certainty there can be that a single elevation value characterizes it within a specified error bound (Dutton and Chan 1983). From a grid, a pyramidal quadtree data structure was created coding each data element with 2 bits, indicating whether the elevations of children of a quadrant moved up, moved down or stayed the same level as their parent. This added two bits to the z-values of quadrants at each level, allowing elevations to differentiate as they multiplied. Thus, by the eighth subdivision (a 256 x 256 grid), elevations were stored as 16-bit numbers, capable of discriminating 64K different z-values. Although it was suitable for interactive zooming of DEMs, the method produced aliasing artifacts in regions of high relief that even a subsequent smoothing of the surface could not eliminate.

An interest in integrating DEM data for arbitrary areas anywhere in the world led to a polyhedral, global data model that employed a variant of DEPTH coding. This was called GEM, for *Geodesic* Elevation Model (Dutton 1984). Never implemented, GEM was intended to enable entry of point elevation data into its store from any source having horizontal coordinates expressed as latitudes and longitudes, integrating these observations into a global DEPTH hierarchy, encoded to a degree specified by a vertical and/or a horizontal error parameter. GEM is a more complex structure than QTM, being based both on an octahedron and on its polyhedral dual, a cube. The octahedron was aligned to the Earth's poles, and scaled to have a slightly larger circumsphere than the Earth. The cube's vertices lie at the centers of the octants, but at a radius that fit inside the Earth by the same distance the octahedron exceeded Earth radius. The vertices of these two figures were connected, yielding 24 initial triangles, edges of which included edges from both polyhedra, as figure 2.9 shows.

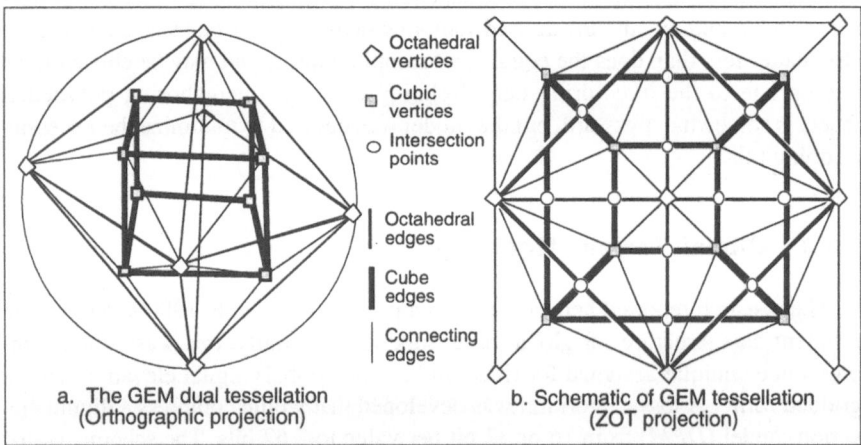

a. The GEM dual tessellation
(Orthographic projection)

b. Schematic of GEM tessellation
(ZOT projection)

Fig. 2.9. Structure of the Geodesic Elevation Model

This Archidemean solid (a rhombic dodecahedron) was recursively subdivided by introducing new vertices in the centers of existing triangles, triangulating them with each other and existing nodes. New vertices "pimple" or "dimple" (move away from or toward the Earth's center, respectively) depending on the elevation values for points connected to them, with the default being not to change elevation. To indicate these conditions, leaf node data describing a facet's elevation change was coded to 01, 10 or 11, respectively.

As figures2.9 and 2.10 illustrate, GEM uses dual hierarchies, but they are not quaternary. At every other level one hierarchy or the other would densify, yielding nine triangular facets for every extant one. Location identifiers would be built up using the numerals 1 through 9, rather than 0 through 3 as QTM employs. As every point on Earth was part of both hierarchies, either one or an interleaving of both could be used to generate *location codes*.

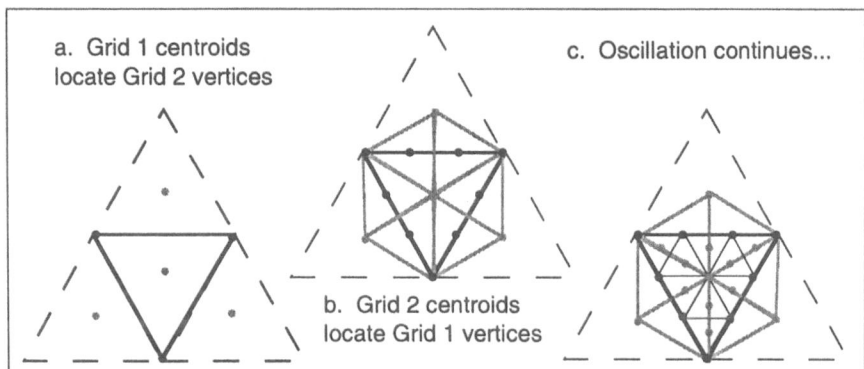

Fig. 2.10. Relations between GEM's dual triangular hierarchies

Although GEM could accept elevation data for any or all points on a planet, it was realized that even a medium-dense elevation model for a whole one would consume huge amounts of storage, even at the rate of less than one bit per triangular facet theoretically possible. Some portions of a GEM pyramid would always be omitted, and some localities would have fewer levels of encoding (resulting in a smaller number of larger facets) than others. GEM's high degree of data compression was based on the implicit ordering of facets in complete sub-hierarchies; the bits in a GEM archive indicated changes in elevation by their values, and identified facets by their ordering, as do elevation grids, only pyramidally. To retrieve elevations across a region, elevations for its quadrants would need to be recursively evaluated (adding and subtracting the delta codes) by traversing the DEPTH pyramid to evaluate the polynomial for each leaf node at the desired level of detail. These elevations would then be assigned to the corresponding GEM vertices in the triangulation defining that level of detail. As GEM's triangulation is fixed and regular, these locations can be generated in parallel to their elevations; being implicit, they do not need to be recorded in a database. Finally, if being retrieved for visualization, the geographic coordinates and elevations would be projected directly or indirectly into display space. For other purposes, a 3D triangular mesh could be generated, and if required, contours and other lines could be threaded through this regular structure.

GEM's chief drawback was that it was never implemented, and thus claims made for it were never properly evaluated. It would have required extremely effective storage management in order to partition sub-trees into pages and access them in an orderly way. Dutton (1984) also claimed that to some extent, a GEM archive could monitor and enforce its own data quality, by rejecting input points whose elevations did not match the coding for (more precise) elevation points already

DEPTH encoded nearby. This assertion presumed that the aliasing problems found in the gridded version of DEPTH modeling could be overcome in GEM.

Lastly, GEM is a complicated computational device, as its spatial indexing reflects two interlocking geometric figures, for which elevations can be described by interleaved digits in their geocodes; each region at a given level in one hierarchy named the nodes of the next level of the other, alternating as they plunge into detail. While each tessellation is space-exhausting, the two overlap, and need to be intersected at some level of detail in order to obtain a combined space partition. All this is possible to compute, but seems excessively complicated for purposes of location coding unless the benefits of dual tessellation clearly outweigh their costs.

2.3.4   Related Independent Research

Entirely coincidentally, a global model very similar to QTM was developed at NASA Goddard Space Center around 1990. The *sphere quadtree* (SQT) of Fekete (1990) recursively decomposes facets of an icosahedron into four triangular "trixels" to spatially index geodata (satellite imagery) and manipulate it directly on the (approximated) sphere. No attempt to encode cartographic data into SQT was made in the manner described here for QTM. The addressing scheme for SQT is like that for QTM (using digits 1-4 instead of 0-3), but starting with an icosahedron results in a forest of 20 quadtrees rather than eight. As a result, relations between individual trees occur more often and are more complex than those in QTM, involving more rotations. Algorithms were devised for finding neighbors, connected components, and other basic tasks, but SQT apparently was never developed any further. Figure 2.11 compares Fekete's SQT and Goodchild and Shirin's HDSH tessellations. Both are quaternary triangular meshes, but derive quite different characters from the polyhedra that root them.

In the field of computer aided design and computer graphics, Schröder and Sweldens (1995) describe using *wavelets* (Chui 1992) to describe functions and distributions (such as height fields or reflectance functions) on a sphere. Spheroids to be modeled are parameterized as an octahedron, then subdivided recursively into quadrants. A variant of classical wavelet transforms that employs a lifting scheme over an area (in contrast to the scaling and dilation operations that generate wavelets on the real number line) that is relatively insensitive in terms of approximation error to the lifting method used. The paper describes and illustrates encoding and reconstruction of the ETOPO5 global DEM (resampled from 5 down to 10 arc-minutes) using spherical wavelets; 15,000 wavelet coefficients resulted in ca. 5% total error (noticeable smoothing), while 190,000 coefficients (9 levels of detail) resulted in a 1% error in reconstruction. No compression statistics were provided, but graphic reconstruction of wavelet coefficients was noted as requiring about 10 minutes on a 150 MHz RISC workstation.

2.3.5   Derivative Models and Applications

An early implementation of QTM was done by Goodchild and Shiren (1992) in a hierarchical spatial data structure (HSDS) project undertaken at the National Center for Geographic Information and Analysis (NCGIA) site at the University of California at Santa Barbara (UCSB). HSDS used a different and simpler facet addressing scheme than QTM. Like QTM, the addressing method placed quadrant 0 in the center of each set of children, but unlike QTM the locations of the 1- 2- and 3-quadrants did not permute; 1-codes always identified "top" or "bottom" quadrants, 2-codes always identified "left" quadrants, and 3-codes identified "right" quadrants. However, Shiren developed an algorithm to convert HSDS quadrant codes to QTM codes and back again, using string-substitution rules.

Like Fekete's work at Goddard, HSDS research at UCSB focused on encoding and manipulating data in the multilevel triangular raster defined by the QTM grid. Twelve- and 15-neighbor triangular raster chain encoding methods (further explained in Goodchild et al 1991) were used to identify connected components and enable intersection and dilation of HSDS facets at any hierarchical level. The UCSB team also created *Motif*-based visualization software that let users at workstations to orient a globe and zoom in to regions at specified levels of detail.

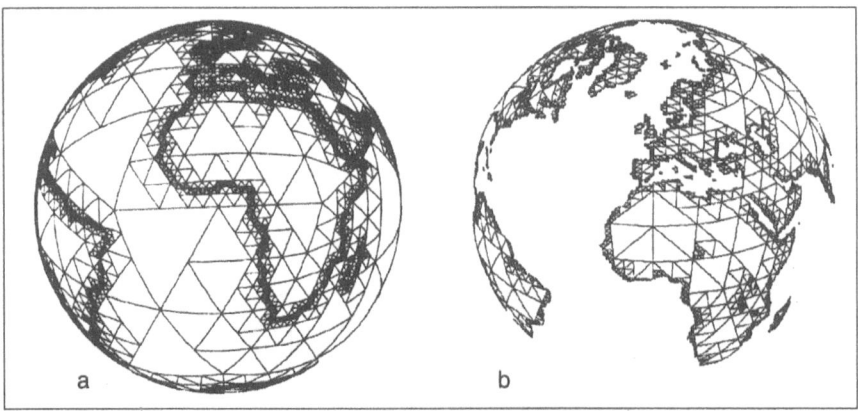

Fig. 2.11. Comparison of SQT and HSDS tessellations. *a*: SQT (icosahedral) rendering of world coastlines; *b*: HSDS (octahedral) rendering of world land masses.

Mentioned above in the context of polyhedral projections, Lugo and Clarke (1995) applied a QTM variant to indexing and compressing digital terrain models. Like NCGIA's HSDS model described above, their triangulated quadtree surface (TQS) utilized the octahedral spherical breakdown and the quartering of facets, but numbered octants and quadrants differently than either the author or Goodchild and Shiren do. TQS Addresses were fixed at 24 bits (12 levels), as the goal was to encode the ETOPO5 global DEM (which uses a 5-minute rectangular grid). Clarke's

Modified Collignon projection (Clarke and Mulcahy 1995) was used to transform the cylindrical ETOPO5 projection into triangular octants arrayed as shown in Figure 2.6. Although no TQS reconstructions of the DEM were presented, and savings in DEM storage appeared not to be very great, the authors did conclude that the method is useful for terrain modeling, particularly because of its capability to retrieve elevations at a hierarchy of scales.

Citing as predecessors both Dutton's QTM and Fekete's SQT, Otoo and Zhu (1993) developed yet another variant of QTM, which was claimed to be optimized for efficient spatial access and dynamic display of data on a sphere, similar to the goals of Fekete and Goodchild and Shiren. A completely different spatial number-ing scheme for facets, called "semi-quadcodes" (SQC) — which Otoo and Zhu claimed to be more efficient in addressing locations on a sphere — was developed for this purpose. These are addresses into a "semi-quadtree," described as the lower half of a rectangular quadtree having triangular quadrants (see figure 2.12). Like the HSDS of Goodchild and Shiren, conversion from latitude and longitude to SQC address employs a rectangular projection yielding an x,y value which is assigned to a bucket at a specified level of detail; a semi-quadcode is then computed for the bucket. No information is available concerning further development of the SQC approach.

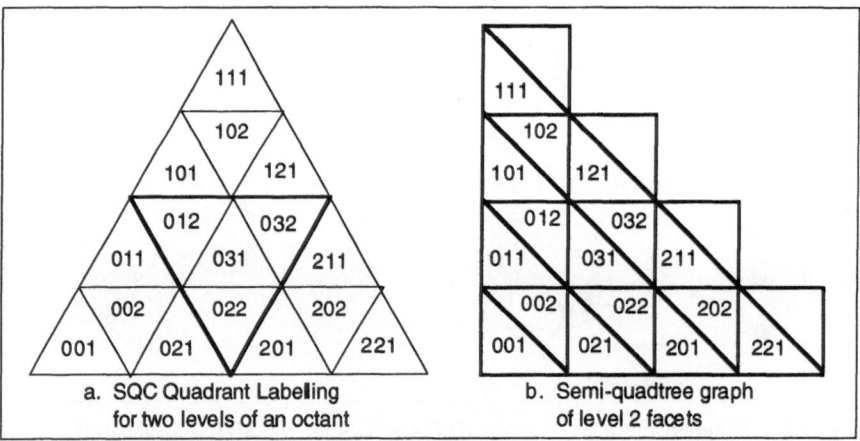

Fig. 2.12. The Semi-Quadcode QTM variant (from Otoo and Zhu 1993)

Lastly — in the celestial sphere — an astronomer at Goddard Space Flight Center (MD, U.S.A.) has suggested using QTM as a possible method by which the Guide and other star catalogues might be spatially indexed (Barrett 1995). The paper compares QTM indexing to *bit interleaving*, also known as *Morton encoding* (Morton 1966), of coordinates of Right Ascension and Declination to achieve angular resolutions of about one millisecond in a 64-bit word. Barrett sug-gests that using QTM addresses for this purpose (looking outward from Earth rather than inward) might provide a more nearly equal tessellation with more inherent contiguity than the cylindrical projection implied by ascension and declination. The issue was not decided in Barrett's paper, which might have referred to NASA co-worker Fekete's SQT model, but didn't.

2.4   Chapter 2 Summary

This chapter has argued in favor of and presented examples of approaches to geospatial data modelling that are able to overcome limitations of both planar and spherical spatial coordinates. There is a need, as Goodchild (1990) has articulated, to be able to model geographic phenomena directly on a sphere, not only in plane projection. Basic properties of a computational model that does this, called a *quaternary triangular mesh*, were described and enumerated. Similar, independent or derivative models were surveyed and some of their applications outlined.

These related research efforts possess certain similarities, despite addressing different sets of applications. All of the authors adopt a global, hierarchical approach, using subdivisions of polyhedra to encode geospatial data. The polyhedra and the forms of tessellation do vary, but not as much as they might. In addition, because they are space-primary models, all these approaches are *regular* (although some are more geometrically uniform than others), and most of them exploit this regularity to perform either point sampling or spatial indexing (further discussed in section 3.1).

Our approach differs from all of these, in that the tessellation is used to build a hierarchical coordinate system for irregular (vector) geometries, as depicted in the diagrams in figure 1.5. This strategy essentially replaces all the coordinates in a spatial database with quadtree addresses (QTM IDs). While this is not in itself a spatial indexing method, it does not preclude also using the model for this purpose. An approach to spatial indexing is explored in chapter 4, exploiting the properties of *nodal regions* called attractors. Before moving on to these concerns, however, the next chapter will examine the computational properties of the quaternary triangular mesh hierarchy, both good and bad, actual and hypothetical.

3 Computational Aspects of a Quaternary Triangular Mesh

We shape clay into a pot,
but it is the emptiness inside
that holds whatever we want.

— Tao Te Ching

As we have seen, there is more than one way to model planets as polyhedra and more than one way to carve up any of these. In general, the models under discussion here start with a platonic solid. Besides octahedra (having 8 facets), we have seen examples of dodecahedra (12), icosahedra (20) and truncated icosahedra (32, itself a polyhedral subdivision). Cubic earth models such as figure 2.4 illustrates have been proposed but will not be dealt with here. To our knowledge, no one has suggested using the simplest polyhedron — a tetrahedron (4 facets) — as the basis for a global spatial data model. Regardless of its orientation, it would generate an even more non-intuitive coordinate system, a stranger distribution of land masses, and introduce more projection distortions than larger polytopes make necessary. But it is certainly possible to do this.

Even in this tidy little universe of solid forms, there are quite a lot of geometric possibilities, each with its own computational twists. A full and systematic treatment of types and applications of planetary tessellations — analogous to works by Coxeter (1973) and Fuller (1982) with respect to polyhedra — would be useful and welcome, but exceeds the focus of this thesis by a large margin. An even larger literature exists concerning planar and higher-dimension tessellations, not only in mathematics, but also in crystallography, computer science and geography as well; discussion of this too must be deferred. So, for better or worse, from here on we shall discuss only octahedral models, and only quaternary subdivisions of these. For the most part, we will dissect the QTM model, referring to properties of related ones when this might clarify key computational considerations without adding excessive entropy.

3.1   Geometric, Topological and Numerical Properties

An overview of the structure of QTM was provided in Chapter 2, section 2, which specified the embedding and facet numbering of the octahedron that is the origin of the QTM hierarchical tessellation (see figure 2.3). Its six vertices, or *nodes*, define three polar axes; each node and axis has a *basis number* which is either 1, 2 or 3. Both the locations of nodes and their numbering is arbitrary, but not unreasonable. Antipodal nodes have the same basis number, such that each octant is defined by a sequence of node IDs that, starting at the North and South poles, are in the cyclic order 1-2-3 either clockwise (in octants 2, 4, 6 and 8) or anticlockwise (in octants 1, 3, 5, and 7). To generate a QTMID for a location, the octant where it falls is subdivided into four triangular quadrants by bisecting each of its three edges and pushing this midpoint upward to the surface of the earth. The edge midpoints form nodes at the new level, and are connected by chords passing through the earth. These chords create four child triangles of slightly different shape and size; the central one always is most equilateral and has the greatest area. Each of these first-level quadrants may in turn be quartered in exactly the same way, such that the number of degrees of latitude and longitude that an edge spans always is divided by two. The higher the level (the more numerous and smaller its triangles are) the closer its facets lie to the surface of the earth and the more nearly co-planar neighboring facets are. Figure 3.1 illustrates the process of facet subdivision for the first two levels. Only the central facets are made opaque, in order to better show geometric differences between levels.

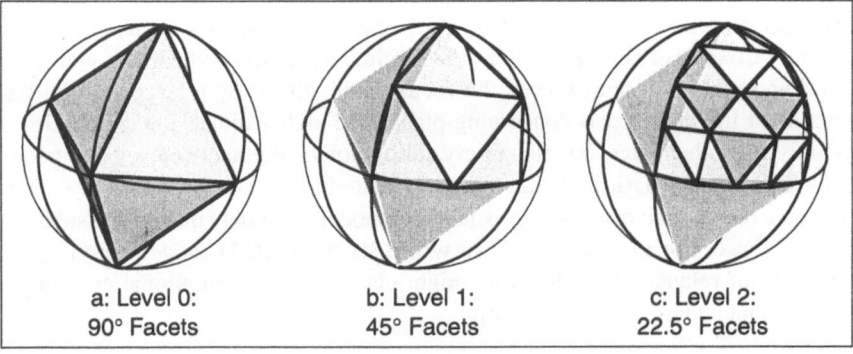

| a: Level 0: | b: Level 1: | c: Level 2: |
| 90° Facets | 45° Facets | 22.5° Facets |

Fig. 3.1: Geometric Development of QTM Levels 1 and 2 from an Octahedron

3.1.1   Variance in Area of QTM Facets

Subdivided facets of hierarchical polyhedral tessellations do not naturally possess equal areas; this only holds for a few basic polyhedra, such as the Platonic solids and some Archimedean ones (which have other attributes, such as unequal angles, that allow this). As Chapter 2 described, equality of areas may be important for certain applications, such as environmental monitoring (Wickman et al 1978; White et al 1992). For such purposes one must either modify the geometry of tessellations or resort to map projections. As QTM is meant to serve as a spatial indexing mechanism and a hierarchical coordinate system, strict areal equality — although it would be welcome — is a requirement it is possible to relax. Rather than glossing this over, however, it seems worthwhile to document QTM's variability, as a time may come when it is necessary to account for it, if not compensate for it in computations. This section gives a statistical summary of the model's geometric non-uniformity.

Doing this turns out to be slightly tricky. As QTM does not follow great circle arcs (beyond those defining a spherical octahedron), it is not easy to compute the spherical areas of its facets. All edges beyond the inital 12 are defined by small circles (east-west edges are parallels, the other two sets are more complex). To get around this, we mensurate the facets themselves, as opposed to the spherical areas they subtend. This procedure computes the chords which connect adjacent QTM vertices, then evaluates the planar areas they enclose, which as tessellation proceeds, lie ever closer to the enclosing sphere, asymptotically approaching it.

We can approach the problem by (1) locating all the points in latitude and longitude that define the vertices of a given level of detail; (2) for each facet in this breakdown obtain the coordinates of its three vertices; (3) compute the three chords given these endpoints (this can be done in two ways: (a) get great circle distance, then chord length; (b) convert to x,y,z, then compute Pythagorean distance in 3-space); and finally (4) compute the area of the facet from the chord lengths. According to Snyder (1987), lengths of great circle arcs can be computed as:

```
cos(arc)     = sin(lat2)sin(lat1)
             + cos(lat2)cos(lat1)cos(lon2-lon1)
```

It is more useful for our purposes to use this form:

```
sin(arc/2) = {sin²[(lat1-lat2)/2]
             + cos(lat2)cos(lat1)sin²[(lon2-lon1)/2]}1/2
```

as this result is half the desired chord factor. Snyder also says that this is more accurate for small arcs. Once the chords of three edges are computed, the planar area of the facet is:

```
Area = (s(s-a)(s-b)(s-c))1/2, where a, b, and c are the three
chords, and
s = (a+b+c)/2   (or half the perimeter).
```

The above formulae presume a unit sphere (having a radius of 1 unit). The following functions compute the quantities involved, using great circles rather than Cartesian products. Note that the great circles are only used to derive chord factors, and are not presumed to be the geodesics that define QTM edges when projected to a sphere. Angles are expressed in radians.

```
real function fArea(lat1,lon1,lat2,lon2,lat3,lon3)
   a = chordLen(lat1,  lon1,  lat2,  lon2)
   b = chordLen(lat2,  lon2,  lat3,  lon3)
   c = chordLen(lat1,  lon1,  lat3,  lon3)
   return triArea(a, b, c)
end fArea
real function chordLen(lat1, lon1, lat2, lon2)
   slat = sin((lat1-lat2)/2)
   slon = sin((lon2-lon1)/2)
   return 2*sqrt((slat*slat)
    + (cos(lat2)*cos(lat1)*slon*slon))
end chordLen
real function triArea(a, b, c)
   s = (a+b+c)/2
   return sqrt(s*(s-a)*(s-b)*(s-c))
end triArea
```

These functions were applied to every facet in one octant over the first five QTM levels. At each one, the computed areas were appended to a list and summed. When all facets were evaluated, the list of areas was sorted and displayed as histograms (figure 3.2). Minimum and maximum facet area, means, standard deviations and kurtosis were computed and reported in table 3.1. The area sum is multiplied by 8 to get a total area for facets of each QTM level, which in the last row is ratioed to the area of a unit sphere. (4π).

Area Measure	Level 1	Level 2	Level 3	Level 4	Level 5
Facets (n)	4	16	64	256	1024
Min Area	0.28974	0.07554	0.01917	0.00481	0.00120
Max Area	0.43301	0.13188	0.03466	0.00877	0.00220
Max/Min Area	1.49448	1.74583	1.80803	1.82328	1.83333
Mean Area	0.32555	0.09346	0.02424	0.00612	0.00153
Median Area	0.28974	0.09342	0.02387	0.00611	0.00152
Std Dev	0.07164	0.01612	0.00354	0.00083	0.00020
SDev/Mean	0.22006	0.17248	0.14604	0.13584	0.13072
Kurtosis	1.74493	2.91530	2.99556	2.44369	2.15422
Total Area	10.41775	11.96281	12.41066	12.52712	12.55654
Sphere Area	12.56637	12.56637	12.56637	12.56637	12.56637
% of Sphere	82.90183	95.19705	98.76091	99.68765	99.92174

Table 3.1: Octant Area Statistics for QTM Levels 1-5 on a Unit Sphere

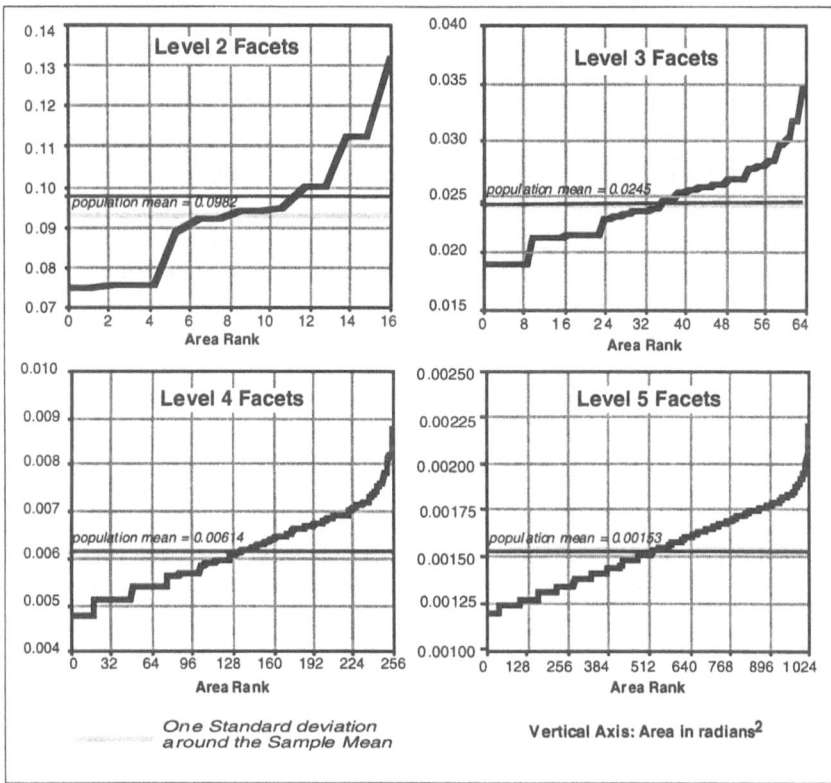

Fig. 3.2: Areas of QTM Facets for Levels 2-5 on a Unit Sphere

Table 3.1 and figure 3.2 reveal several interesting properties of this hierarchy. First, the area ratio of the largest facet to the smallest one at each level approaches about 1.85 (i.e., in the limit facet area varies plus or minus 43% from the average). While QTM is not an equal area tessellation, neither is it highly variable. Note also that the largest facets (those in the central level 1 quadrant nearest a pole) are quite distant from the smallest facets (touching the octahedral equatorial vertices), separated by about 8,000 km. Neighboring facets tend to have equal or nearly equal areas, especially at sizes used to resolve coordinates (level 10 and beyond). Relative areal variability slowly decreases, as indicated by the ratio of standard deviation to mean area; this drops from 0.22 to 0.13 after five subdivisions, revealing that extremes in area become less prevalent at higher levels. This is confirmed by the kurtosis statistics, which decline from values typical of a normal distribution to a level characteristic of a uniform one (Bulmer 1967: 64). Finally, the total area of a sphere is rather quickly approximated; after five subdivisions the summed area of the resulting 8,196 facets exceeds 99.9% of a sphere. We may conclude that QTM generates a relatively uniform partitioning of the globe, although as Chapter 2 stated, it is possible to do better if one needs to by using other tessellation schemes.

These findings demonstrate that QTM's spatial resolution is not constant across a planet; while this is hardly a surprise, it does have consequences. One is that the facet area and resolution column figures in table 2.1. fib a bit, as areas actually vary by close to a factor of two and linear resolution by up to 40% within any particular QTM level of detail. This could have subtle effects when employing QTM detail filtering to cartographic data, as is described in Chapter 4. For example, areas in extreme latitudes — such as Canada or Scandinavia — would receive slightly more generalization than areas nearer the equator, such as West Africa or Southeast Asia, due to the fact that quadrants extending from mid-latitudes toward the poles are somewhat larger than quadrants that lie near the equator, especially in the vicinity of the four octahedral vertices that occupy it.

3.1.2   QTM Quadrant and Node Numbering

We number QTM octants from 1 to 8, while their quaternary subdivisions are numbered from 0 to 3. Why not number the octants from 0 to 7 and save one bit per QTM ID? The answer depends on the internal representation of QTM IDs, which may be either integers, bit fields or strings. It also matters whether the length of QTM IDs (the number of levels represented) is fixed or variable. While every coordinate in an entire dataset could be encoded to a uniform precision, there may be times when this condition should be relaxed. Indeed, the ability to specify how resolution or accuracy varies is one of the chief benefits of using this encoding to begin with. This section describes intrinsic properties of QTM numerology, then delves into reasons behind them and some consequences of these choices.

Figure 3.3 summarizes the first two levels of detail. Note that grid nodes are not given unique IDs, only basis numbers. We will describe in section 3.2 several approaches to uniquely identifying nodes, as they play important roles in QTM's structure and applications. Corner and outer edge nodes in figure 3.3 are aliased, a consequence of using the ZOT projection in this diagram (see Appendix A for a description of ZOT computations). Starting with the octahedral vertices (level 0 nodes), level 1 nodes are developed at edge midpoints and numbered 1, 2 or 3, according to the formula:

$$B_n = 6 - (B_a + B_b) \qquad \text{(Dutton 1989a: 135)}$$

where B_n is the basis number of the new (midpoint) node, and B_a and B_b are existing nodes from the prior level that define the edge being subdivided. This computation is performed by a function (*QTMeval*, described in Appendix A) that generates QTM IDs from geographic coordinates and vice versa. At each subdivision of a facet, a computed node number is appended to the existing QTM ID, except where the location being encoded occupies a central quadrant (shaded in figure 3.3) at the current level of detail. Central quadrants are always assigned the digit 0, and are not identified with a mesh node (they identify parents' centroids). All other (non-central) quadrants are identified with a particular 1- 2- or 3-node, and therefore all quadrants incident to a given node share its basis number as their last digit. Except at the six octahedral vertices (where four quadrants meet), each node has six incident quadrants, none of which have the same parent quadrant (but usually have earlier common ancestors); they are all cousins rather than siblings. This can be verified by inspecting quadrant QTM IDs in figure 3.3.

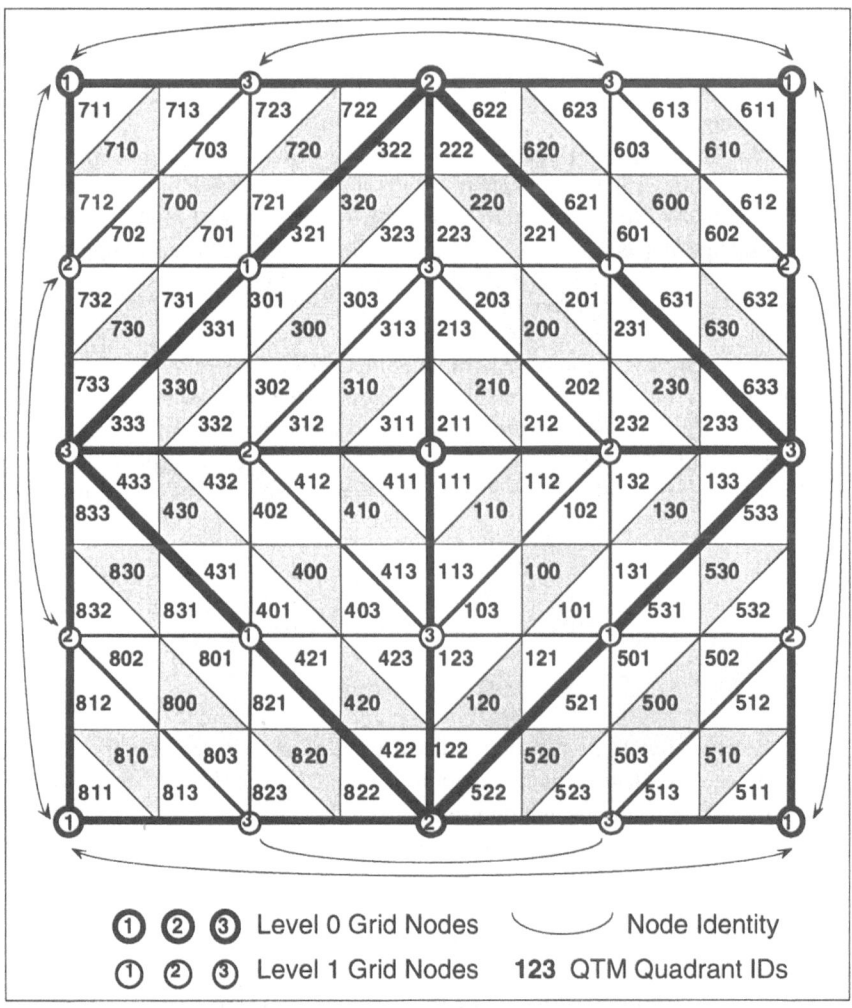

Fig. 3.3: QTM node and facet numbering to level 2, ZOT projection

The method of numbering QTM quadrants and nodes just described is not the only possible approach. As sections 2.3.4 and 2.3.5 summarized, other facet numbering schemes for triangular quadtrees are in use that seem to work equally well. The spherical models described by Fekete (1990), Goodchild and Yang Shirin (1992), Otoo and Zhu (1993) and Lugo and Clarke (1995) all handle these matters in slightly different ways, resulting in labels for facets that are not the same as ours or each other. Of these studies, only Fekete's provides for explicit node numbering, used for identifying facets that share a common vertex. Goodchild and Yang's work also includes ways to identify facets surrounding a node, but this is done by string substitution of quadrant IDs, on the fly, without ever referring to nodes by name. Section 3.2 will further articulate the need for node identifiers, and describe several approaches to formulating them. Chapter 4 describes possible roles for QTM nodes, in the context of map generalization via QTM hierarchical detail filtering.

Octant ID Properties. Recall that we number octants from 1 to 8, proceeding clockwise in the northern hemisphere from the prime meridian, then continuing in the same direction in the southern hemisphere. As described in (Dutton 1996) a simple function that computes octant numbers from geographic coordinates is:

```
OCT(lon, lat) = (1 + lon DIV 90) - 4 * (lat - 90) DIV 90
```

Negative longitudes must be complemented prior to computing octants. The function will yield an incorrect result at the South Pole but nowhere else.

Also, given the ID of an octant, it is easy to identify neighboring ones. North, South, East and West neighbors meet along octant edges, except for the North neighbors of octants 1-4 and the South neighbors of octants 5-8, which meet only at the poles:

```
EAST_OCT(oct)  = 1 + (oct + 8) MOD 4 + (4 * (oct DIV 5))
WEST_OCT(oct)  = 1 + (oct + 6) MOD 4 + (4 * (oct DIV 5))
NORTH_OCT(oct) = 1 + (oct + 9 - 2 * (oct DIV 5)) MOD 4
SOUTH_OCT(oct) = 9 - (oct + 9 - 2 * (oct DIV 5)) MOD 4
                   - (2 * oct MOD 2)
```

The last two functions can be modified to eliminate vertex neighbors should there is no need to identify such transitions. Similar functions have been developed to identify QTM quadrants neighboring a given one. These are more complicated, due to the permutations involved in numbering quadrants.

3.1.3 QTM Coordinate Conversion

Any point location on earth can be represented as a QTM facet or node identifier, at any spatial resolution desired. The greater this resolution, the greater the number of digits in the point's identifier. Our implementation limits this number to 29 levels of detail (hierarchical subdivisions), as this amount of information will fit into a 64-bit binary word. It allows for a global spatial resolution of ca. two cm, which — as section 3.1.1 documents — varies across the globe, but by less than a factor of two. In this section we describe encoding latitude-longitude tuples into QTM IDs, as well as the inverse process, recovering geographic coordinates from QTM identifiers. We shall see that the former is deterministic, and can be as exact as source data permit; the latter, however, is non-deterministic, but is capable of generating coordinates that are within the error limits of the source data, hence statistically indistinguishable from those locations.

As implemented in this project and documented in Appendix A, computing a QTM ID from a point location is a simple iterative or recursive procedure. The method used generates QTM quadrant vertices as well as QTM IDs. When encoding locations, facet vertex locations may be superfluous to the user, but for decoding (recovering a point location from a QTM ID) they are required by a subsequent step in which one point within the facet is computed (or one of its corners selected) to approximate the desired location. Either way, a single — but critical — parameter is needed, one which differentiates QTM from other forms of coordinate storage. This, an *accuracy criterion* (a distance tolerance), specifies how well locations are known. Typically it is expressed in *meters on the ground*, and halts both encoding and decoding by limiting the number of QTM levels to compute. Facet subdivision ceases when the edge length of the next more detailed level would be less than the specified limit of accuracy.[1] As *QTMeval* is called every time a point location is encoded or decoded, its accuracy parameter can change whenever it is necessary to adjust spatial resolution.

Two stages of encoding a point location are illustrated in figure 3.4. The edge size of facets, in this diagram, is compared to the tolerance criterion in order to halt encoding. The figure reflects the method employed by *QTMeval* to identify quadrants using the ZOT projection. How often it is necessary to change this criterion is totally data- and application-dependent. When encoding geodata into QTM coordinates, the accuracy/error of the source should normally dictate what distance tolerance is specified to *QTMeval*. This can be constant for a file, a feature class or a feature, and rarely might vary within features. However, when encoding computed geodata that result from merger, overlay or other analytical operations, the locational accuracy of features can become heterogeneous. If it is possible to document the lineage of each location presented to *QTMeval* for encoding, an appropriate tolerance for it can be specified, and the resultant QTM IDs will reflect that knowledge. We call this capability to completely specify locational variability *building in positional metadata* (see figure 1.5). It is the heart and soul of our claim of QTM's superiority over other geographic coordinate systems.

---

[1] This test is actually performed in ZOT space, along the X axis (the Y axis has the same scale), by *QTMeval*, using a tolerance parameter (in ZOT units) passed to it. Refer to Appendix A for further discussion and listings of *QTMeval* and related algorithms.

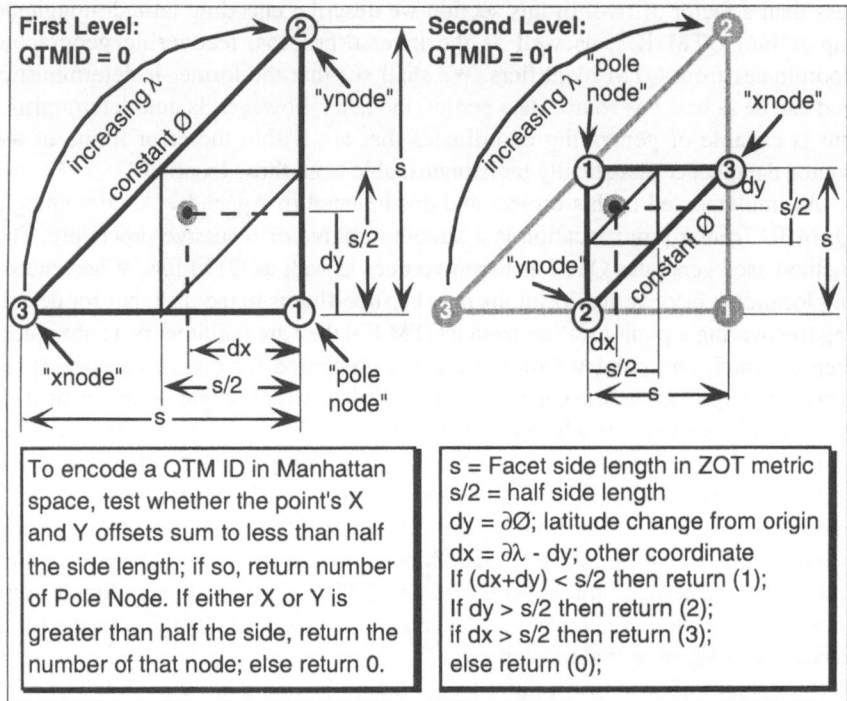

To encode a QTM ID in Manhattan space, test whether the point's X and Y offsets sum to less than half the side length; if so, return number of Pole Node. If either X or Y is greater than half the side, return the number of that node; else return 0.

s = Facet side length in ZOT metric
s/2 = half side length
dy = ∂Ø; latitude change from origin
dx = ∂λ - dy; other coordinate
If (dx+dy) < s/2 then return (1);
If dy > s/2 then return (2);
if dx > s/2 then return (3);
else return (0);

Fig. 3.4: Two stages of encoding a point location (adapted from Dutton 1991a)

Converting a QTM ID back to geographic coordinates is a parallel exercise to encoding it, but doing this involves additional considerations that are application-dependent rather than data-dependent. As QTM data specifies how well each location is known, a limit is placed on the available accuracy; this comes from positional metadata doing its thing. But the purpose and scale of coordinate retrieval is entirely a function of the application, and this can change the values of coordinates that are generated when QTM IDs are decoded under program control. When decoding a QTM ID, an application must make two basic decisions:

1. Choose precisions at which to decode locations (up to the limit stored for them)

2. Choose positions for decoded locations within facets

The first choice determines the spatial resolution and/or map scale of output coordinates. Whenever such retrieval is done at a level less than the encoded precision, the possibility exists that more than one coordinate will occupy the same quadrant. If the coordinates represent a point set or a *polyline*, this indicates that point filtering is required to reduce quadrant occupancy, generally to a single vertex per quadrant. Which "competing" vertex get selected, however, is up to the application; several approaches to doing this are discussed in chapter 4.

The second choice is necessary because any location within a selected facet is as good as any other at that facet's resolution. Five general possibilities exist:

1. Select a random location inside the facet;

2. Use the centroid of the facet;

3. Select one of the three vertex nodes of the facet;

4. Compute a location weighted toward one vertex;

5. "Micro-position" the point using additional detail stored for it.

These alternatives are diagrammed in figure 3.5.

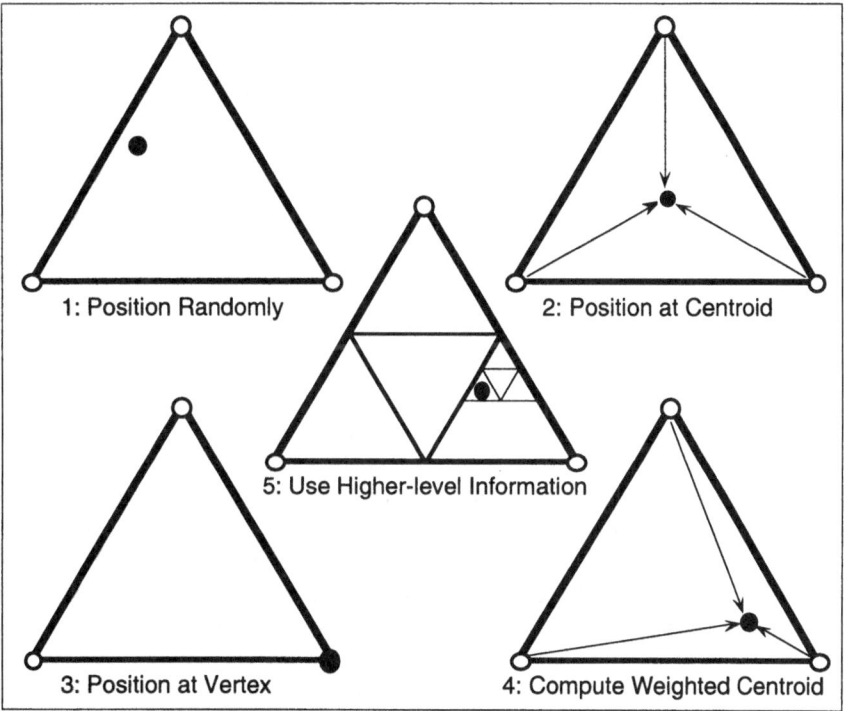

Fig. 3.5: Five Options for locating coordinates decoded from a QTM ID

The nature of the application may dictate which approach is used, but such decisions must be well thought-out, and can change from time to time. Method one (random positioning) is usually appropriate only for point observations, and then mostly for statistical sampling purposes. It provides a way to generate stratified random samples, if this is needed. The second method is a maximum-likelihood approach that is often appropriate for polyline data, as it yields consistent, unbi-

ased point positioning. Method three places point locations at nodes of the triangular mesh (this only works for QTM IDs terminating in 1, 2 or 3; when using it,
method two is usually applied to IDs that end in 0). Doing this causes points that
occupy adjacent facets to be placed at common locations. This approach may be
used as part of a more elaborate filtering process, also described in chapter 4. The
fourth option is a generalization of the second and third ones; a weighted centroid
is computed that is biased toward one vertex, usually the one indicated by the last
digit of the QTM ID. It can be used to cause points to gravitate away from areas of
congestion or toward one another without actually colliding. Finally, method five
(Dutton and Buttenfield 1993) can be used to position points close to their encoded
locations, even though they are being selected (filtered) at a coarser resolution.
Using it prevents the point locations from wandering and assuming excessive regularity at coarser levels. The full encoded precision of each QTM ID can be used to
compute a geographic location, or any degree of detail in between that and the
level at which retrieval is being performed. Using method five still requires selection of one of the other four methods for computing the point's actual location
(using one of the first four approaches); choosing the centroid of higher-level quadrants (method two) is least biased, and was used in the applications reported in
chapters 4 and 5.

3.1.4   Structuring QTM Identifiers

Previously it was mentioned that our implementation of QTM allows for 29
hierarchical levels of detail. One of the reasons given was that this amount of information will fit in 64 bits, the same amount of storage occupied by a double-
precision floating point number on most computing systems. The prior statement
assumes that octant codes (which vary from 1 to 8) will occupy 4 bits, and will be
followed by 29 or fewer quadrant codes, each occupying 2 bits. This actually adds
up to 62 bits; the remaining 2 bits may be used in various ways or ignored. However, it may not always be most efficient to use a binary representation for QTM
IDs. One reason is that few computing platforms and programming languages
support 64-bit integers, which is the primitive datatype that would normally un-
derlie this abstract one. To make code more portable, one might be able to alias
QTM IDs with double-precision floating point numbers (e.g., using a C language
union or a Fortran *equivalence*). However, in ANSI C, bit fields are limited to 32
bits, and Fortran does not support them as a primitive datatype.

For these and other reasons, most software that handles such identifiers tends to
represent them as *strings*, which wastes memory and disk space, but makes it
more efficient for lower-level routines to process location codes. When QTM IDs
have fewer than 16 digits, string representation may not waste too much space (no
more than double precision coordinates do). In our current testbed, however, string
IDs are internally allocated at maximum precision (30 characters) to make it easier
to vary precision when necessary. When writing a file containing QTM IDs to

disk, any trailing unused characters in them can be stripped out, and run-length compression techniques can reduce storage space even further. But it would be much more efficient to use a binary encoding for this purpose, and convert QTM IDs to strings when processing specific objects extracted from the file. This would use an amount of secondary storage for encoded geodata equal to what single-precision (x,y) floating point coordinates would occupy, or half that for double-precision ones, while providing increased information content.

Enriching Identifiers. A 64-bit fixed-length binary format for QTM data was proposed by Dutton (1996). In it, IDs would be right-shifted to place any unused bits at the high-order end of words, and these would be set to zero. On the average — data encoded from a typical 1:25,000 scale map — there will be about 20 bits not required to store IDs. The higher the precision of source data, the more quaternary digits it will take to encode its coordinates, and the fewer "leftover" bits will exist in QTM identifiers. An approach to utilizing as many such unencoded bits as may be available for storing additional metadata or attributes is presented in (Dutton 1996). Inserting bit fields in unused portions of IDs can characterize particular locations in qualitative or quantitative fashion in order to assist applications in making processing decisions. Figure 3.6 illustrates this "enrichment" concept and provides coding examples of types of metadata that might be included in such bit fields.

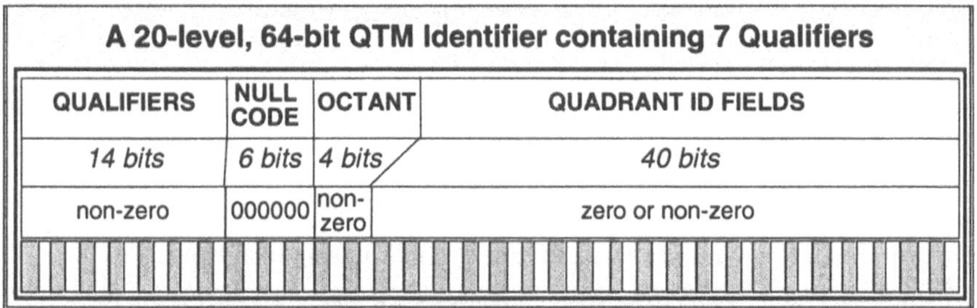

A 20-level, 64-bit QTM Identifier containing 7 Qualifiers			
QUALIFIERS	NULL CODE	OCTANT	QUADRANT ID FIELDS
14 bits	6 bits	4 bits	40 bits
non-zero	000000	non-zero	zero or non-zero

Some possible locational Qualifiers and encoding conventions

QUALI-FIER	SUBJECT MATTER	Code_10	Code_01	Code_11	PROPERTY CONVEYED
1	Vertex Type	Node	Point	Mixed	Topological role
2	Vertex Origin	Defined	mputed	Mixed	Locational importance
3	Vertex Angle	Sharp	Smooth	Mixed	Local angularity
4	Feature Genus	Point	Line	Area	Topological dimension
5	Feature Origin	Cultural	Natural	Mixed	Source of entity
6	Feature Width	Fixed	Variable	Mixed	Uniformity of extent
7	Feature Priority	High	Low	Medium	General malleability

Fig. 3.6: Enriched QTM ID 64-bit binary word structure

Encountering qualifiers such as the above can trigger or parameterize application actions. For example, in the context of feature generalization, Dutton (1996) suggests that hints provided by qualifiers can be used in the following ways, among others:

• If successive locations along a feature are coincident, merge the points

• If one coincident location is less precise than the other, move or delete it

• If one coincident location is a node, move or delete the other one

• If one location is on a smoothed *primitive*, adjust it first

• If one feature is cultural and the other natural, adjust the cultural one

• Operations should preserve linear feature widths and/or separations

• Operations should respect feature class priorities

Enhancing Identifiers. Rather than including extra bit fields that need separate interpretation, it is possible to "enhance" a QTM ID by *appending one or more extra digits to it*. When coded as zeros, such digits do not change the location of the point, only its purported precision. One reason for doing this is to indicate that a point should receive special treatment by a generalization operator. This assumes that there is a "normal" precision for the point, one that it and other points in the same primitive or feature received during their initial encoding. Appending zero bits to any point in an object can flag that location for reasons the application determines and presumably is able to interpret. For example, a vertex where a polyline turns at a right angle could be considered more significant than neighboring ones where inflection is not rectilinear, but this can depend on the type of feature the polyline represents (such a rule would only be applied to man-made features).[2] Enhancing that point's precision makes it possible to retain the vertex when performing a subsequent feature simplification. An application can vary the number of appended zeros to indicate how "important" a point has been judged to be (by prior analyses) for changing map scale and filtering line detail.

Likewise, one could mark points as being less important, hence as candidates for deletion, by dropping trailing QTM ID digits (effectively replacing them with zeros). This could also be used, for example, to give one feature or feature class lower priority than others. It can also be combined with enhanced precision to provide heuristics for both point selection and deselection. Such tampering with vertex precision would normally be only temporary, and needs to be used judiciously to avoid losing information or propagating false precision in a database. In our testbed, we did not implement enhanced QTM IDs as described above. Instead, we allocated two bit fields in each ID to store heuristics for vertices. That strategy is described in chapter 4 and appendix A.

---

[2] This would be roughly equivalent to setting qualifiers 2, 3 and 5 to '10' in an "enriched" QTMID, as illustrated in figure 3.6.

3.1.5   Spatial   Access   Strategies

As the discussion on spatial data models in section 2.1 described, spatial data tend to be structured either regularly or irregularly, flat or hierarchically, and globally or locally. Virtually all *regular* data models employ *space-primary* structuring, while *irregular* models utilize *object-primary* structuring. For example, a space-primary digital elevation model (DEM) is normally constructed as a regular grid of cells all the same size (either geographically or when projected), which provide implicit location coding according to their ordering. In such models, objects can only be identified via attribute coding; this can be done using separate overlays to specify grid cells occupied by each object type (e.g. a hydrography layer that codes only water areas). Object-primary DEMs also exist, and are most often modelled as triangular irregular networks (TINs). In a TIN, triangle vertices tend to be placed at critical surface points (peaks, pits, passes, pales and breaks in slope), and their edges tend to follow critical surface lines such as ridges, valleys and shorelines. The idea is to preserve the structure of the landscape as accurately as required, without over-specifying the surface, as regular grids always do. While the "objects" that are modelled in TINs are not what a lay observer would necessarily notice or be able to name, from a modelling perspective their triangles are all specific (albeit connected) entities capable as being identified and manipulated individually, or as a neighborhood constituting a terrain feature.

In the field of spatial data handling, the space-primary/object-primary distinction has held fast for many years, and has usually been described as a *raster/vector* dichotomy (Bruegger 1994). In the raster world, location is the primary access key, while in the vector world, object identifiers are primary. To identify and access objects in raster data requires searching through pixels to link them. This operation is often called *connected component labelling*, as it identifies and labels as one component all grid cells that are adjacent and share certain attributes.

Most GISs use specialized data structures (enhancements to the basic vector-topological data model) and algorithms to perform *spatial indexing* for each coded object. A spatial index is similar to an index in a tabular database that identifies a specific record from one of its attributes (such as a date, an invoice number or an employee ID) without having search through an entire set of records. However, as spatial data are ordered in at least two dimensions, and most spatial features are *extended* objects (occupying regions of space rather than points), indexing them is much more difficult. The problem of mapping data in two or more dimensions to one-dimensional lists of sequential memory addresses has many solutions, and some of them have found their way into commercial database systems, such as Oracle (1995), described below.

As QTM and related models are space-filling tessellations, they provide a sound basis for spatial data indexing. A simple example of how well QTM works for the purpose of indexing point data is provided in appendix E. It shows that QTM is a relatively efficient indexing method for global data, although further research is needed to determine strategies that will allow extended objects to be indexed as efficiently as point data.

Efficient, rapid access to vector GIS data via location requires a robust spatial indexing strategy, which needs to be well-adapted to the content and structure of databases, application objectives and computing environments. Access most often tends to be in the form of *range queries*, such as "find all portions of interstate highways that pass through Madison County," "identify all parcels that abut the Shawsheen River" or "how many people live within 10 km of a nuclear facility?" To answer such queries, three stages of computation are generally necessary:

1. The spatial extent of the query must be computed;

2. Indexes of database objects within the range are computed or identified;

3. Records for the objects satisfying the query may then be fetched.

Additional sub-stages of retrieval may occur; for example, many indexing schemes match range queries against rectangular bounding boxes of objects (which may be stored as R-trees or related access mechanisms), and then must perform further analysis to determine if the object itself satisfies the query or only its box does. QTM encoding permits this for features, and also defines minimum bounding triangles that can also be used to initiate spatial searches.

Even though commonly thought of as a specialized form of raster data model, quadtrees may be used to spatially index vector data. For example, Chen (1990) outlines a method for using a quadtree (rectangular and planar, but of unspecified type) to spatially index triangles in a TIN, in order to provide efficient responses to range queries. This method was implemented as an access mechanism in the Arc/Info™ GIS TIN module. Oracle Corporation offers a "spatial data option" for its Oracle7™ Relational Database Management System (RDBMS) that has a great deal in common with the present approach, but is limited to indexing (as opposed to integrating and displaying) multi-dimensional data (Oracle 1995). Users identify the dimensions (database column) to be indexed —for example, latitude, longitude and date — which are then automatically scaled and combined into a hierarchical index, called an HHCODE. These codes may be recursively decomposed into nested subsets of data space, to a user-specified limit of resolution. If the resolution is relatively coarse, many data items may map to the same HHCODE; the RDBMS automatically partitions tables when many records map together, and allows users to extend HHCODEs to optimize indexing and accommodate additional data records (Oracle 1995: 16).

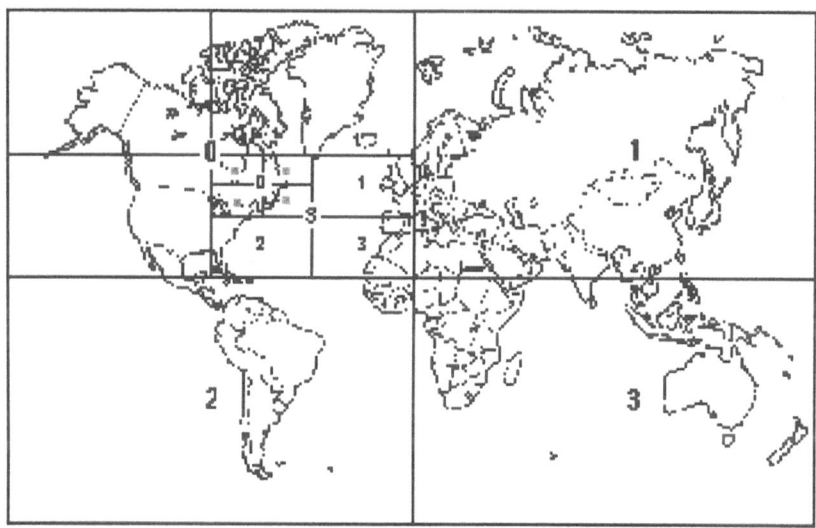

Fig. 3.7: Quadtree global decomposition for HHCODE spatial indexing
(from Oracle 1995: 5)

Figure 3.7 is reproduced from an Oracle white paper. The document does not state if this particular (unspecified) map projection is used to build the quadtree hierarchy for geographic HHCODEs. Note that the projection is not a cylindrical one, does not appear to be equal-area, does not partition along the equator or prime meridian, and appears to contract the Atlantic ocean while expanding the Pacific ocean. From this we assume the map is only a conceptual illustration.

3.2   Attractors: Knitting Hierarchies Together

Using QTMIDs in place of coordinates models point locations as specific, triangular areas, thus providing a basis for handling the uncertainty and scale of point measurements. Partitioning space into triangles, while computationally convenient, yields a somewhat artificial domain for modelling positional uncertainty, which is usually described in terms of circular or elliptical deviations around a point. But as the essence of a hierarchical coordinate system is *approximation of location*, the shape of elements used to model locations should not be of critical importance, especially if child facets properly nest within their parents without gaps and overlaps. Error approximation might be more precise if hexagonal tessellations were employed (because hexagons are more circular), but hexagonal hierarchies do not nest properly, and in any case it is impossible to tessellate a sphere using only hexagons (but see White et al 1992).

As section 3.1.1 shows, QTM's tessellation is not geometrically uniform (no hierarchical subdivision of a sphere into triangles can be), but this is not a critical weakness. Perhaps its biggest shortcoming — which it shares with all other quadtree data structures — is that spatial relationships between facets are only explicit between children and parents and between siblings. That is, given a leaf node (QTM ID), IDs of parents and grandparents are easily determined (they are initial substring), and the IDs of siblings are obvious (they have different terminal digits), but the identities of other neighboring or nearby facets are less simple to discover. Neighboring facets may in fact have quite different IDs, especially if they lie across a major hierarchical partition. For example, two very close points, one slightly below the 45th parallel, the other slightly above, might have QTM IDs that differ only in the 2nd digit, for example 1133333333 and 1033333333. In a conventional quadtree, it is possible that these data would be stored in memory or disk locations that are quite far apart, depending on how sub-trees are spatially indexed. Using QTM as a hierarchical coordinate system — as we do — complicates this further, as the same QTM ID may occur in a number of features which may need to be handled together. GIS feature data may be well-indexed in secondary storage, but this does not assure that nearby (or even identical) point locations will be stored near one another, regardless of the scheme.

A hierarchical approach to relating proximal neighbors was proposed at the time QTM was originally described (Dutton 1989a). It exploits the dual relationship between QTM facets and nodes to associate the facets surrounding each node in the global mesh. These neighborhoods are *hexagonal* groups of six facets interspersed with *triangular* ones. These regions (plus the six *square* neighborhoods around octahedral vertices) completely partition the sphere . At each level of subdivision, these regions densify; new ones sprout in new locations and existing ones generate children that occupy their center point but have half their diameter. We call these nodal regions *attractors*, as one can think of them as points with large masses that exert gravitational influence on all locations in their well-defined neighborhoods. Figure 3.8 presents an idealized view of four levels of attractors, showing their constituent triangular quadrants. Only a few QTM facets, hence attractors, are as equilateral as this figure shows; most are skewed, such as the ones shown in figure 3.9. Triangular attractors are left unshaded, allowing larger ones below them to be glimpsed.

Attractors may be used to partition map space at a resolution appropriate to the data and map resolution being worked with. For example, a map generalization application may select a set (and size) of attractors that relate to a particular target scale, and use them to identify the features that fall within each one in a systematic fashion. Using this approach, attractors are not stored as part of a database; rather they are dynamic data structures called upon as needed to identify spatial proximities and potential conflicts in specified localities, and can then be dismissed.

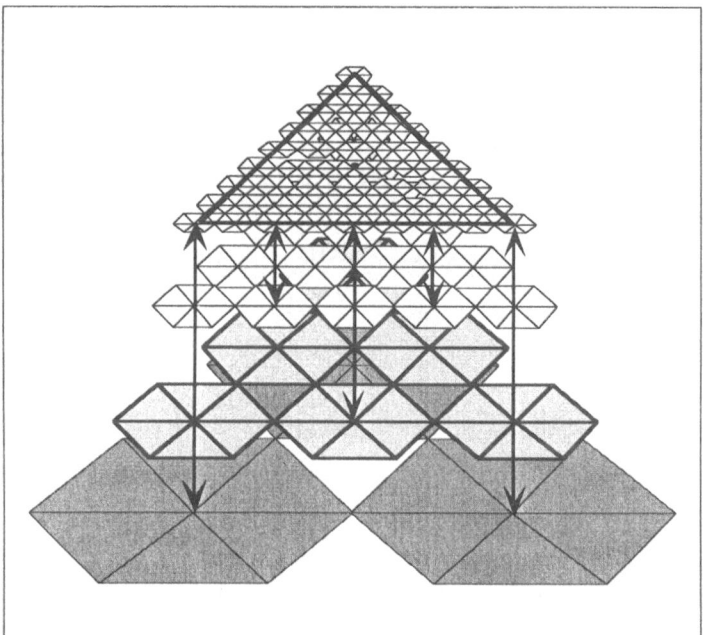

Fig. 3.8: Four levels of attractors occupied by a QTM facet;
multiple ones exist everywhere except at the central point

3.2.1   Attractor Formation and Naming

Each attractor has exactly one "proper" (concentric) child, and covers sectors of
six "improper" (eccentric) children that have a second hexagonal parent that touches the first one at a point, plus two other potential parents. For any given eccentric child (always hexagonal) any of four lower-level attractors can be considered
parents: two hexagonal attractors centered at each endpoint of the *segment* being
split, and either of two triangular attractors that also touch at that midpoint. In a
quadtree, each parent has four children, and each child has one parent; Here, each
parent has one proper child and six improper ones, and each child may have up to
four parents! While this is a kind of inheritance hierarchy, attractor relations are
better modeled as a directed acyclic *graph* (*DAG*) than as a tree.

This complex of relationships is what makes attractors useful for associating
point locations: the hexagonal ones unite points in neighboring sub-trees of the
QTM hierarchy. The triangular ones do not do this, but serve as keys to identifying their three hexagonal neighbors; as these are the siblings of the triangular attractor, all have the same ID except the least significant digit (*lsd*), which will be
1, 2 or 3 (IDs of triangular attractors all end in 0). Given a QTM facet belonging
to a hexagonal attractor, the attractor's identifier (called an *AID*) can be computed

as the numerically smallest QTM ID of the six included facets. As an example how such naming can operate, table 3.2 presents AIDs for 11 levels of attractors at two nearby locations in northern Switzerland, starting at QTM level five.

Location A		Location B	
LAT,LON: 47.60452, 8.507163		LAT,LON: 47.601349, 8.521080	
QTM ID: 113301310010*1311223*		QTM ID: 11330131001000*210130*	
LVL	**AID**	**LVL**	**AID**
5	113321	5	113321
6	1133323	6	1133323
7	11330131	7	11330131
8	113301310	8	113301310
9	1133013100	9	1133013100
10	11330131121	10	11330131121
11	113301310010	11	113301310010
12	1133013100*121*	12	1133013100*100*
13	11330131001*313*	13	11330131001*212*
14	113301310010131	14	113301310010131
15	113301310010*1311*	15	11330131001*00210*

Table 3.2: Attractor Identifiers for a pair of nearby locations in Switzerland

In table 3.2, differences between AIDs are highlighted. Note that the common ancestor in the sequence is 1133, only four digits long (a spatial resolution of more than 600 km). From level 11 (5 km resolution) on, all the attractors share the base 11330131001, which has 11 common leading digits. Observe the abrupt shift in AID at level 10, then back again at level 11. Also observe that at level 14, the same attractor occurs, but appears to descend from two different parents. These two points lie 1.1 km apart, which is within the size of a level 14 attractor (1.22 km major diameter). They happen to be placed near the center of attractor 113301310010131, and their parents were adjacent attractors that had a common vertex at the center of it. The level 14 attractor itself does not, strictly speaking, have a proper parent; it blossomed from a point (the common vertex of two level 13 attractors) at that level of detail.

Notice the general similarities between attractor and facet identifiers in this table. This results from using QTMIDs to name attractors. In the case of triangular ones, AIDs are simply quadrant QTM addresses. In the case of hexagonal attractors, the lexically smallest QTM ID of the six facets in each set is selected to identify their attractor. Under this naming convention, AIDs are simply facet IDs, to which zero or five other neighbors are aliased (or only four, in the six special cases of cardinal-point octahedral vertices).

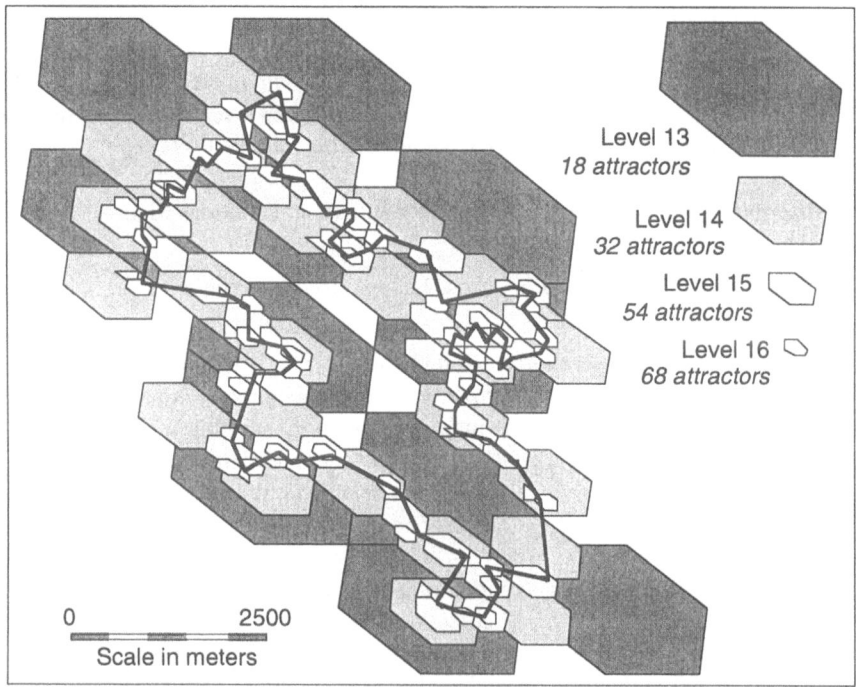

Fig. 3.9:   Four levels of vertex attractors for a polygon with 74 points

Only a minority of attractors have a proper parent. Because of how they are
spawned, one cannot simply remove lsd's (unless they are repeating digits) from
AIDs to ascend their hierarchy. Therefore, it is not as easy to associate attractors
from different levels as it is for QTM IDs. That is, larger attractors underneath a
given one may share none, some or all of its leading digits (but usually have some
of the most significant ones in common, as the above example indicates). The
general frequency of occurrence of "orphan" attractors is indicated by the map of attractors shown in figure 3.9. In that figure, the only attractors shown are those
that this polygon's vertices occupy. Most are hexagonal, but several triangular
ones also occurred. The number of attractors at each level is a general indication of
the potential for reducing line detail when implementing some of the map generalization techniques discussed in chapter 4.

3.2.2   Attractors as Containers

An application program can choose when, where and which levels of attractors
to evaluate as part of its geoprocessing strategy. But when it chooses to do this,
what exactly should happen? Having named an attractor occupied by a vertex, this
information needs to be put somewhere, at least temporarily. The main purpose of
doing this would be to "collect" the vertices that "belong" to a given attractor. As

the same or other features are processed and additional vertices fall into an already-identified attractor, this collection will grow — not without limit, but arbitrarily large. The amount of data an attractor may hold grows in direct proportion to data density, and also will tend to be larger for bigger (lower-level) attractors than for smaller ones. There are two design decisions that this raises:

1. What is the minimum set of data items that need to be stored when a vertex is added to an attractor?

2. How can these groups of observations be organised for efficient search, update, data retrieval and deletion?

These questions will be addressed in chapter 4. The first one can only be answered by considering the role of attractors in one's geoprocessing application. In this thesis, the application focus is map generalizaton, and this will color subsequent discussion. As our purpose in using attractors is to identify feature elements that are competing for map space, we need to be able to trace back to these objects from the attractors they occupy. The second question is a spatial indexing problem. It is not obvious what — if any — existing method for allocating storage for simple, ephemeral, numerous dynamic objects such as attractors is the best approach. Linked lists, heaps, hash tables, trees and directories are all potentially useful techniques for this, but novel data structuring approaches may be called for. Section 4.2.3 looks at these issues and evaluates alternative approaches.

3.3   Documentation and Discovery of Positional Accuracy

To encode geographic coordinates into QTM IDs, two types of information are needed. First a latitude and a longitude (or plane coordinates for which an inverse projection is known) must be provided for point locations. In addition, some measure of the accuracy or certainty of these locations is required. This can come from metadata, from standards, or be estimated by users, but it must be supplied, and in a form that enables a ground resolution to be computed. As chapter 1 discussed, such error terms — when available — are likely to apply to an entire dataset, or at least to entire feature classes or layers. If a spatial dataset has complicated lineage, such error parameters may no longer be valid, or may vary spatially and thematically within it.

QTM encoding is highly capable of modelling variations in spatial accuracy, at the dataset, feature class, feature, primitive and vertex levels of detail (see figure 1.5); the encoding algorithm (see appendix A) simply needs appropriate error data for every vertex it handles. This may not always be easy to provide in such disaggregated form, but if points are encoded with incorrect accuracy parameters, some benefits of using QTM may be diminished or lost. After an examination of method-based error, this section discusses how QTM can help to estimate positional error for coordinates of unknown or untrustworthy accuracy. This approach has the interesting side effect of generating estimates of fractal dimensions (Mandelbrot 1978) for each feature and/or feature class analyzed.

3.3.1   Method-induced Positional Error

Before discussing how positional error in map data can be estimated, an examination of errors inherent in QTM's structure and algorithms are necessary. As the analysis reported in section 3.1.1 showed, the QTM grid exhibits areal variations across the globe, which never exceed a factor of two and vary minimally when transiting from one facet to the next. Another type of error not discussed above, but potentially having greater impact on data quality evaluation, is the *positional error resulting from inexact transformations into and out of the QTM realm*.

Table 2.1 indicates the general magnitude of QTM's locational uncertainty in its *Linear Extent* column. This represents the average mesh size (distance between adjacent quadrant vertices at each level of detail), and does vary across an octant by about plus or minus twenty percent. Facets are smallest near the equator, larger near octant centers, and largest near the poles. Whether such variations are significant depends upon one's application; it certainly would complicate computations of distances between facets, were they to be done in the QTM coordinate system, but we do not do this.

When a point location is encoded into a QTM ID, it can be made just as precise as the information warrants, yielding a certain number of digits in the ID. When that ID is decoded back to latitude and longitude coordinates — even if all its precision is utilized — there is still a residual uncertainty about where to actually place the point within the last facet reached. The least biased estimate results from using that facet's centroid as the point location. The amount of error in placing it there will be less than the point's inherent positional error, assuming all QTM digits are used in decoding the location. But how much error might be expected, and how might it vary systematically?

Because a series of transformations are required, including projection (to ZOT coordinates) and deprojection (to spherical coordinates), all utilizing floating-point arithmetic, an analytic answer to this question is hard to formulate. The amount of residual error depends also on how these transformations are computationally implemented. Given these factors, an empirical investigation seemed necessary to verify that residual errors lie within reasonable bounds. Such a study was performed, and its results are reported in the remainder of this section.

To verify that the QTM encoding and decoding algorithms used in this project (see appendix A) function within their expected error bounds, sequences of points were generated across the full extent of one octant (as positions in all octants are computed in the same way, only one octant was analyzed). Points were first sampled along meridians at a spacing of 0.60309 degrees, yielding 149 samples of 149 points (22,201 data points). Each data point was transformed to a QTM ID at level

20, which has a nominal resolution of 10 meters. It was felt that this (medium-scale) precision would suffice to indicate positioning accuracy. Each data point's ID was then decoded back to a latitude and longitude at 10 meter precision, and the resulting coordinates compared with the input values. Coordinate data was stored in 32-bit floating-point numbers, and double-precision arithmetic was used in making comparisons. Spherical great circle distance between respective input and output coordinates was then computed, and the data tabulated for analysis.

The same procedure was then repeated, this time sampling along parallels, starting at 0.60309° spacing. To compensate for degrees of longitude not being constant, the number of samples along each parallel was scaled by the cosine of latitude; this resulted in 149 runs of observations, which began with 149 sample points at the equator and decreased to only one near the North pole, totalling 36,377 observations. As the QTM grid is defined by doubling parallel spacing, and samples were taken across a non-rational interval, this caused a systematic drift of observations toward and away from the QTM East-West edges, and thus periodic variation in the amount of error from the equator to the pole. The same effect occurs — but to a barely noticeable degree — in runs of samples taken along meridians. The effect is minimal because the azimuths of non-East-West edges of the grid vary, and run directly North-South only along edges of octants.

Table 3.3 and figure 3.10 show the results of this analysis. Note that the mean errors in the two directions are very close — less than 3.3 meters, well under the 10-meter resolution of QTM level 20, and also less than its 5-meter Nyquist wavelength. As expected, deviation about the mean is greater along parallels than along meridians, by a factor of about 8. The two curves in figure 3.10 graph the errors of recovered positions, averaged across each of the 149 transects in either direction. Virtually all of the variation seems to be noise — beat interference with the QTM grid. However, the longitudinal error appears to decline slightly at mid-longitudes, while latitudinal error slowly rises in moving from the equator to the pole. The former effect may reflect the greater compactness of triangles near the centers of octants. The latter trend is probably due to the increasing size of quadrants as they approach the poles.

Transect	Latitudinal	Longitudinal	Combined
No. Samples	14 176	22 201	36 377
Min. Error (m)	1.38334	2.80093	1.38334
Max. Error (m)	6.46030	3.71752	6.46030
Mean Error (m)	3.18627	3.28127	3.24425
Std. Dev. (m)	0.86617	0.10951	0.48784

Table 3.3: QTM Level 20 Positional Error Statistics for Transects

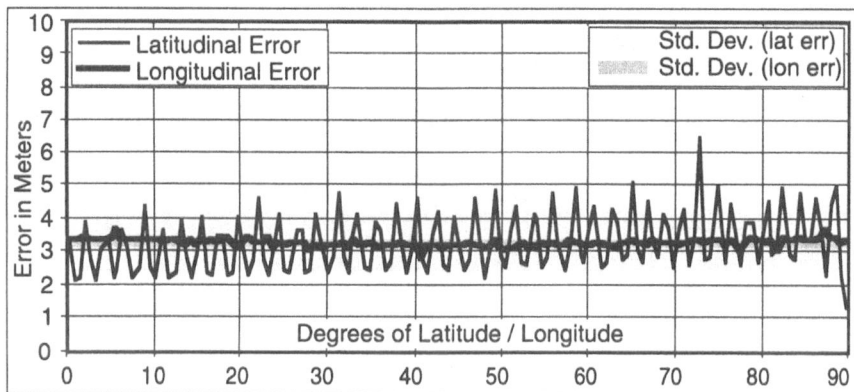

Fig. 3.10: Systematic Error from QTM Point Decoding at Level 20

The effects of quadrant size and shape hypothesized above are further substantiated when the positioning statistics are broken down by level 1 QTM quadrants. The two sets of measurements are classified by quadrant in table 3.4. Quadrant 1 (which touches the North pole) has the greatest area and exhibits the largest average error, followed by quadrant 0 (the central one below it). Equatorial quadrants 2 and 3 are the smallest, and have the least error. In addition, within-quadrant differences between the two sample sets are smaller (0.62% maximum) than the overall difference (3.0%), further indicating that quadrant size and shape (rather than algorithmic errors or sampling biases) most likely accounts for the slight trends noticeable in figure 3.10.

Quadrant	Lat. Err. (m)	N (Lat)	Lon. Err. (m)	N (Lon)
0	3.14013	3 638	3.15975	4 355
1	3.47030	4 110	3.47432	11 026
2	3.02696	3 214	3.06333	3 418
3	3.03460	3 214	3.03009	3 402
ALL	3.18627	14 176	3.28126	22 201

Table 3.4: Mean Level 20 Positional Error by QTM Level 1 Quadrants

3.3.2 Unearthing Positional Error

When digitized boundaries of unknown accuracy or scale are encoded into QTM hierarchical codes, statistics describing reconstructions of them at different levels of detail can indicate both the upper and lower limits of useful detail. Such analyses, which can be made visually or numerically, indicate what might be an appropriate positional error parameter to apply to the data for encoding. Superfluous digits in the data's QTM ID strings can then be removed from the encoded data with confidence that no significant line detail is likely to be eliminated.

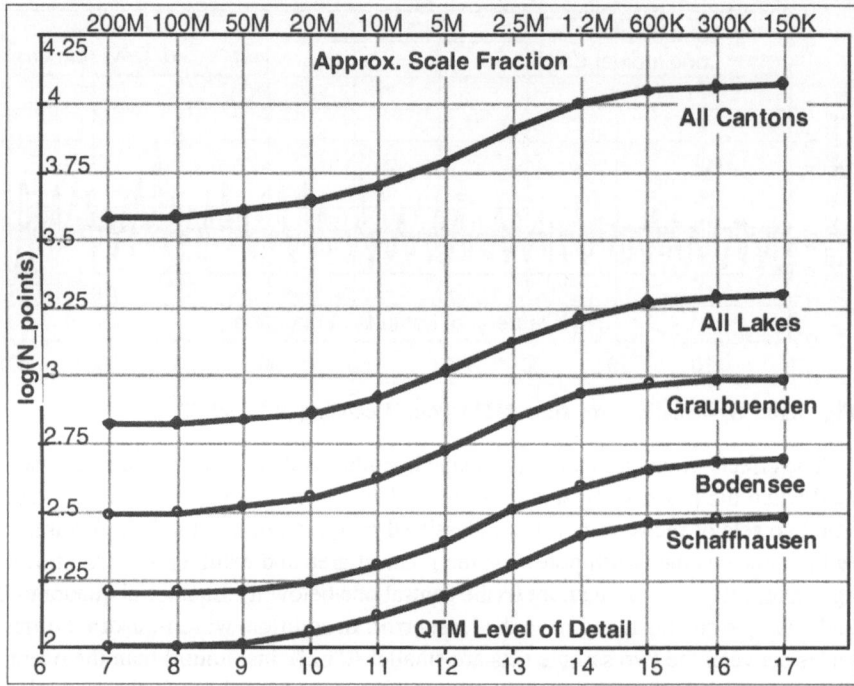

Fig. 3.11: Total number of vertices for Swiss canton boundaries by QTM level

Such an analysis is illustrated in figure 3.11, which reflects QTM processing of Swiss canton and lake polygon boundaries (see figure 3.12) across 12 levels of detail. This particular dataset had a long, not fully documented processing history, that included deprojection from the Swiss national grid coordinate system following (Douglas-Peucker) line filtering using an unrecorded bandwidth tolerance. By filtering QTM coordinates at different resolutions and tabulating how many points were retained at each level, a better picture emerged of the limits of resolution of this dataset. The two top curves in figure 3.11 tabulate sets of polygons; the higher one is the set of all land boundaries, and the one below it includes only lakes. In general, the lake boundaries are smoother (less crenulated) than those for cantons, but their behavior — as well as that of several individual polygons graphed below them — is quite similar across the scales. The similarity may in part be due to the use of the Douglas-Peucker line simplification algorithm, using a constant tolerance for all boundaries. Figure 3.12 shows a map of this dataset, using an equi-rectangular projection. Note how similar many of the boundaries are to one another in terms of their complexity; this probably accounts for the goodness-of-fit of statistics shown in figure 3.13 below.

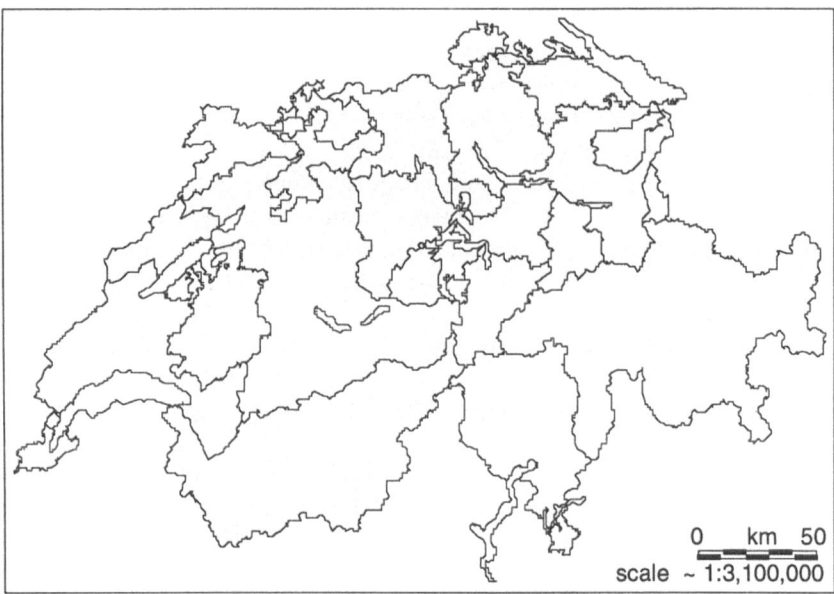

Fig. 3.12: Swiss Canton and lake map data (before conversion to QTM)

The most interesting result that figure 3.11 shows is how the amount of boundary detail levels off both at small scales (to the left) and medium scales (to the right). The flattening occurs at the same place for most of the subsets, below QTM level 10 and above level 15. These resolutions correspond roughly to map scales from 1:20,000,000 to 1:500,000, using the 0.4 mm criterion discussed in section 2.2. This is what we regard as the useful range of resolution for analysis and display of this data, and indicates that no more than 16 or 17 digits of QTM precision are really required to encode it. It could well be that the locational accuracy of the source data is greater than what this implies (QTM level 17 has a ground resolution of about 75 m.), although subsequent processing may have degraded it. Because detail was deleted from the boundaries prior to QTM encoding, it is not possible to say exactly how precise the remaining points are. However, we are able to identify a precision for encoding the pre-processed data which captures all its detail. Thus we have recovered useful positional metadata from undocumented coordinate information.

3.3.3 Revealing Fractal Properties

The same information plotted in figure 3.11 can be transformed to a related form from which we can measure the complexity of individual features (in this case polygonal ones, but in the normal case one would probably study boundary arcs), sets of them or a dataset as a whole. We sum the lengths of segments comprising a feature to get its boundary length. As table 2.1 lists, each QTM level has a characteristic linear resolution. In that table only a global average is provided for each level, but resolution does not vary locally by much. By comparing the lengths of features to the global average resolution by level, it is possible to estimate the complexity of boundary lines. When we do this comparison in logarithmic coordinates, a straight line of negative slope is the expected result if features tend to be *self-similar* (i.e., they look about as complex at large map scales as they do at small scales). The steepness of the slope is a measure of line complexity, and is related to *fractal dimensionality* by the linear equation $D = 1 - S$, where S is the log-log slope parameter and D is an estimate of fractal dimension.

A fairly comprehensive study of self-similarity and fractal dimensionality of both natural and cultural map features was published by Buttenfield (1989). She used a modified divider method based on the Douglas-Peucker (1973) line simplification algorithm, such that step sizes varied both within and between features. Buttenfield claimed that despite these variances, the estimates were robust. The first estimate of D using QTM was published by Dutton and Buttenfield (1993), but these results were based on a detail filtering algorithm that was not as well-behaved as the ones developed more recently. Figure 3.13 presents a fractal analysis of the data we have been looking at.

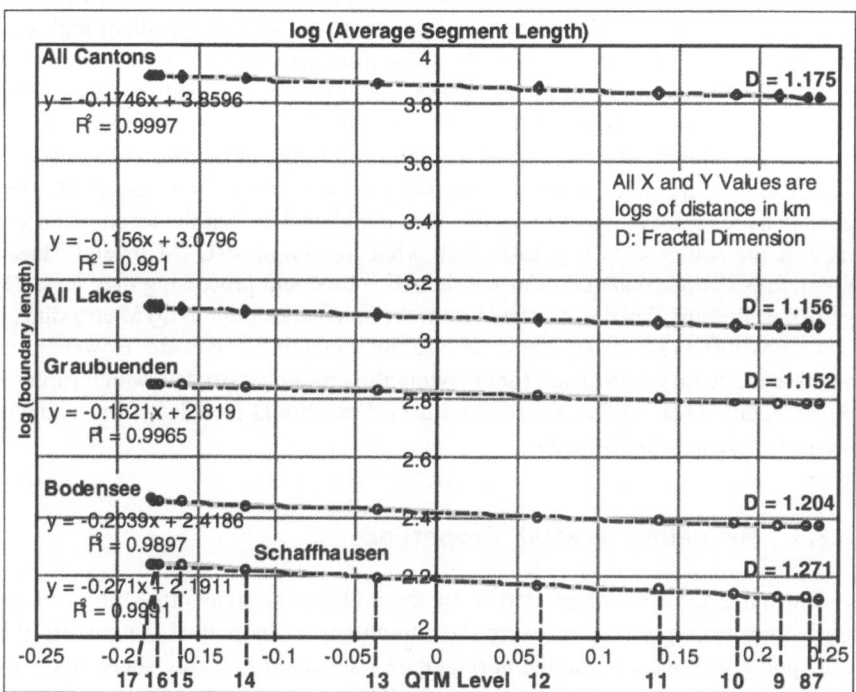

Fig. 3.13: Fractal analysis of Swiss canton boundaries ("Richardson diagram")

Plots such as figure 3.13 are called *Richardson diagrams* after L. F. Richardson, whose empirical studies of coastlines and national boundaries established the "divider method" of analyzing the complexity of irregular lines (Richardson 1961). This is a manual cartometric technique for length estimation in which a divider tool is walked along a line from beginning to end; the number of steps taken, multiplied by the distance between the divider's points yields an estimate of line length. By setting the step size successively smaller, a series of increasingly accurate estimates is built up, which for all curved lines grow greater as the aperture of the divider shrinks.

In Richardson analyses, plots of smooth (e.g., circular) curves stabilize immediately, then at large divider settings drop off suddenly, but the plots of irregular curves behave somewhat differently. For many natural map features, when total length (transformed to logarithms) is plotted as a function of divider aperture, a straight line that trends slightly downward can often be closely fitted to the data points. The relationship between step size and measured length becomes increasingly inverse (its slope decreases) as lines grow more crenulated and detailed. Although Richardson himself was uninterested in the linearity of his diagrams, Mandelbrot (1982) demonstrated that the slope of such regression lines estimates a fractal dimension, assuming that the line being measured is self-similar.

The very strong goodness-of-fit of the regression equations in figure 3.13 indicates high self-similarity of this set of features, simplified by QTM detail filtering. This is revealed through a hierarchical tessellation that itself is self-similar: QTM is a fractal. This implies that it is possible to increase the degree of self-similarity of features through generalizing them in a way that increases the regularity of point spacing (e.g., method 2 in figure 3.5), if this is regarded as useful. Still, this process can subtly distort features, making them look more gridded than they really are. Control over such typification could be exercised through selection of a technique used to compute point locations when transforming QTM IDs back to geographic coordinates, but we did not pursue such possibilities in the map generalization studies reported in chapters 4 and 5.

Figure 3.5 showed five alternative strategies for coordinate conversion, some of which are more self-similar than others. Selecting a centroid or vertex to locate the output point yields the greatest regularity, hence self-similarity. Choosing a random point yields the least, and the weighted centroid method gives intermediate amounts of self-similarity. The fifth option, decoding at a more refined (higher-level) point position, can achieve self-similarity to the degree that the data itself exhibits this property. In sum, any latent self-similarity can be preserved when generalizing QTM data, and more can be added at will.

3.4   Chapter 3 Summary

After describing the structure of the global quaternary triangular mesh tessellation, specific aspects of its geometry, topology, and node and facet numbering were elaborated. Variability of facets was discussed, using an analysis which shows that the largest ones at any level have less than twice the area of the smallest ones, and that these extreme cases are thousands of km apart. Algorithms for computing facet areas were presented, showing summary statistics of variation in area for five levels of detail.

The method of facet numbering was described next, and shown to derive from the properties of node *basis numbers* (either 1, 2 or 3) and how they permute at successive levels of detail. The numbering scheme is compared to other similar models that were described in chapter 2. A set of in-line functions for computing octant numbers from coordinates and identifying neighboring octants were also specified.

Coordinate conversion is the key operation to the aspects of geoprocessing reported in this chapter. A method for deriving QTM IDs from geographic coordinates was presented, as was its inverse, which uses the same algorithm but requires making certain assumptions, as generating geographic locations from QTM IDs is non-deterministic. It is stressed that reasonable accuracy estimates for points are just as essential as their locations to encode them into QTM properly.

The nature and structure of QTM identifiers was described, discussing both string and binary representations. Strings are more wasteful of space, but are more efficient to parse. Binary QTM IDs hold more information in less space (8 bytes versus up to 32 for strings) and are better suited for archival use. Two metadata encoding schemes were presented for "enriching" and "enhancing" QTM IDs; these add useful information in unused portions of the identifiers, either permanently or temporarily.

The next section discussed general requirements for accessing coordinate data. Some access mechanisms that use quadtrees to perform indexing were discussed, noting that although this is facilitated by QTM, the use of QTM IDs for storing hierarchical coordinates does not dictate a particular spatial indexing strategy.

The notion of attractors was explored in the next section, These are nodal regions that are either single or groups of six QTM quadrants, serving to relate data points that fall in neighboring sub-trees. Attractors are identified with QTM mesh nodes, and can be named for one of the quadrants they contain. The hierarchy of attractors was described, some of their roles in generalizing map data were discussed, material which will be amplified in the following chapter.

The chapter concluded with a discussion of how QTM can be used to verify, discover or estimate positional accuracy and certainty of data it encodes. This capability is vital when spatial data to be encoded is poorly documented or heterogeneous. Statistical trend analysis of lines filtered via QTM can reveal the upper and lower limits of useful resolution for files, feature sets and individual features. It was shown that a side-effect of such analyses can result in robust estimates of fractal dimensionality for feature data.

4 Using QTM to Support Map Generalization

We think in generalities, but we live in detail.
— Alfred North Whitehead

Our life is frittered away by detail ... Simplify, simplify.
— Henry David Thoreau

Because map generalization is so often necessitated by changing the scale of representation of spatial data, it is a natural application for a hierarchical coordinate system such as QTM. Not only are QTM-encoded coordinates intrinsically scale-specific (within a factor of two), the partitioning of the planet that the tessellation defines provides a matrix of locations within and among which spatial data can be referenced and manipulated. First this chapter will discuss what "generalization operators" are and which of them might be aided by spatial encodings such as QTM. Differences between hierarchical and non-hierarchical operators will be outlined, to help explain why hierarchical coordinates need not be generalized hierarchically.

After a brief overview of the breadth of operators that QTM might enable or assist, the chapter narrows its focus to a single one, line simplification. A large number of algorithms for this purpose have been proposed and implemented, several of which are briefly discussed to illustrate their properties, strengths and weaknesses. Each of these algorithms requires specification of one or more parameters that control and constrain its operation, values for which are not always easy to decide upon. The discussion of setting parameters leads to the introduction of QTM-based line generalization, an approach to which is then outlined (with further details and algorithms provided in appendix A).

One of the most troublesome and difficult-to-implement aspects of digital map generalization — detection and resolution of spatial conflicts — is addressed next. We describe aspects of conflict identification involving using QTM attractor regions to index areas where potential for congestion or overlap exists. After discussing several alternative strategies for this, our implemented method — a directory-based approach — is described and illustrated.

The final section concerns some useful and important refinements to our basic line simplification algorithm involving vertex selection and line classification. A method for characterizing line sinuosity and embedding this information as positional metadata is described. Whether or not this particular classification method is optimal in all cases, the strategy is a general one that can utilize other parametric methods for steering point selection.

Having detailed our generalization strategies in this chapter, we shall devote chapter 5 to reporting empirical tests of QTM-based generalization operators that reveal some of the behavior of the parameters that affect the process. In that chapter, the system assembled to be the testbed will be described, including the hardware and software platforms involved and the data model employed to structure test data, as well as the datasets themselves. Further tables and figures describing the test runs are provided in appendices C and D.

4.1   Map Generalization: Methods and Digital Implementations

The need for map generalization is as easy to understand as its methods are difficult to formalize. Before grappling with the latter, consider the former from the point of view of a poet and an acute observer of nature:

> Wherever there is life, there is twist and mess: the frizz of an arctic lichen, the tangle of brush along a bank, the dogleg of a dog's leg, the way a line has got to curve, split or knob. The planet is characterized by its very jaggedness, its random heaps of mountains, its frayed fringes of shore. (Dillard 1974: 141)

As for modeling such a mess, Annie Dillard goes on to say:

> Think of a globe, revolving on a stand. Think of a contour globe, whose mountain ranges cast shadows, whose continents rise in bas-relief above the oceans. But then: think of how it *really* is. These heights aren't just suggested; they're there ... It is all so sculptural, three-dimensional, casting a shadow. What if you had an enormous globe in relief that was so huge it showed roads and houses -- a geological survey globe, a quarter of a mile to an inch -- of the whole world, and the ocean floor! Looking at it, you would know what had to be left out: the free-standing sculptural arrangement of furniture in rooms, the jumble of broken rocks in a creek bed, tools in a box, labyrinthine ocean liners, the shape of snapdragons, walrus. (Dillard 1974: 141)

Deciding what to leave out from such a globe (or map), and where and how to do that is the essence of map generalization, a graphic art that has resisted many attempts to verbalize and codify its practice. Although digital mapping has been used since the 1960's and interactive map generation practiced since the mid-1970's, it was only toward the end of the 1980's that systematic overviews of generalization of digital map data were put forward (Brassel and Weibel 1988; McMaster 1989), enabling software engineers to more precisely emulate what map makers do.

Human cartographers generalize maps by imploding graphic representations of places and things on the earth's surface; in the process they decide what original entities to include, and whether and how their symbolization should be altered. Mapping houses have developed guidelines that specify how symbolism should change according to scale, which feature classes take priority over others, and other rules of a more aesthetic nature, such as for managing feature density, label placement and line simplification. In digital mapping environments, which nearly all mapping houses now utilize, such guidelines are still useful, but are not usually specific or formal enough to enable automation of the process, unless a human is constantly available for steering it — the "amplified intelligence" approach to generalization (Weibel 1991).

The tasks that generalization involves have been classified a number of ways. One of the most widely-used typologies is the one developed by McMaster and Shea (1992). In it, about a half-dozen *generalization operators* are defined, each of which may be applied in a variety of ways, depending on the type of map features involved, and via a variety of algorithms, each having its own logic, efficiency, advantages and disadvantages. However, the McMaster and Shea typology — now the dominant paradigm — failed to include an important operator, *feature selection*. Selection (and its inverse, *elimination*) is usually performed first, before any map data or symbolism is changed. Adding this to the list provided by McMaster and Shea and reorganizing it somewhat results in the following set of operators:

- *Elimination/Selection* — determining which features (or feature classes) are significant at a given scale and selecting those to be displayed.
- *Simplification* — Caricaturing shapes and reducing data density in either the geometric or the attribute domain.
- *Smoothing* — Making lines and regions less angular.
- *Aggregation* — Merging adjacent symbols or attribute categories, sometimes changing their form of symbolization.
- *Collapse* — Reducing feature dimensionality; e.g., transforming regions into lines or points.
- *Displacement* — Moving map symbols to alleviate crowding.
- *Enhancement* — Adding (invented, presumed or actual) detail to features to locally or globally exaggerate their graphic character.

A GIS may use a number of different abstract datatypes to encode map features. Different data structures such as polylines, polygons, splines, grids, triangulations or hierarchical data models tend to have specialized generalization requirements, so that the above operators work differently in various domains. Applying them may also cause side-effects, such as altering feature topology, and may require referring to the attributes of features, or even altering them in the course of generalizing map data. Table 4.1, adapted from Weibel and Dutton (1998), gives a simple summary of such contingencies, with illustrative examples of how operators apply to the principal datatypes that GISs manipulate.

OPERATOR	GENERIC DATA TYPE				
	Points	Lines	Areas	Fields	Hierarchies
Eliminate Select	*Eliminate via attributes or context*	*Remove minor branches and boundaries*	<u>Eliminate small areas or sub-polygons</u>	*Alias less significant values*	Prune trees to have more uniform detail
Simplify	Cluster closest neighbors to new points	Eliminate minor segments or inflections	Remove <u>islands</u> and concavities	*Collapse category definitions*	Evaluate at lower levels of detail
Smooth	Make point distribution more uniform	Soften angles, adding small segments if necessary	Minimize or smooth small crenulations	*Average or convolve local variations*	Average at facets or nodes of tessellation
Aggregate Amalgamate Merge	*Combine similar neighbors*	<u>Simplify intersections of linear elements</u>	<u>Delete edges between similar features</u>	Interpolate to larger cell size	Derive lower levels from higher ones
Collapse	<u>Replace by area symbol or convex hull</u>	<u>Combine nearly parallel features</u>	<u>Erode to point, region or medial axis</u>	*Merge similar categories*	Redefine or reorganize hierarchy
Displace	Disperse from each other and larger objects	Increase separation of parallel features	Move away from linear elements	Anamorphose (rare)	<u>Move data to less crowded neighbors</u>
Enhance Exaggerate	Impute new points to densify distributions	Impute plausible characteristic minor details	Complicate boundaries, <u>impute lakes and islands</u>	*Emphasize differences, equalize histogram*	Extrapolate to higher levels of resolution

Table 4.1: Generalization operators applied to types of spatial data. Underlined items may require/result in topological transformations; italic items require reference to attributes or can change their domain. Entries are examples only, and are not intended to be exhaustive.

As table 4.1 illustrates, map generalization has many components and contexts. Most discussions of generalization operators are limited to vector primitives, and do not include the categories *Fields* and *Hierarchies*, as we do here. While these are higher-level constructs than *Points*, *Lines* and *Areas*, we have included them in this table to show that most operators are still valid and in many cases can be implemented with little or no more difficulty than the basic spatial primitives require. In addition, it should be clear that many of the operators in table 4.1 may interact with one another in various ways, which must be dealt with in any map production environment, even for interactive work. Most of these complexities must be put aside here, for better or worse. In the following discussion of QTM-based generalization strategies, primary attention will be paid to the process of *simplification*; in the empirical tests described below and documented in appendices C and D, only results of *simplification of linear and polygonal features* are reported. However, there are other generalization operators that QTM can help to implement, and some of these applications will be mentioned in the next section and discussed further in chapter 5.

4.2 Aspects of QTM-based Generalization

A Quaternary Triangular Mesh is a hierarchical tessellation that divides (geographic) space into roughly similar two-dimensional cells (*quadrants*, which we also identify as *facets*). As these cells all derive from the uniform faces of an octahedron we imagine to be embedded in the earth, and given the method by which they are generated through subdividing existing ones, each of them has a fully-determined (but variable) size, shape, orientation and location on the planet, as chapter 3 described.

It is natural to think of this arrangement as a type of raster, and imagine allocating data to its triangular pixels arrayed across a sphere or spheroid. Indeed, all the related research into hierarchical planetary tessellations described in chapter 2 has taken a raster approach in modeling linear, areal and point-sampled data (Fekete 1990; Goodchild and Shirin 1992; White *et al* 1992; Otoo and Zhu 1993; Lugo and Clarke 1995; Barrett 1995). But just as there are many forms of quadtree data structures, specialized in order to model different types of spatial data (Samet 1990), there are a variety of ways in which QTM and its relatives can also be used.

The principal approach taken in this research is to utilize QTM as a *hierarchical coordinate system* for vector-encoded spatial data, and to explore how this can facilitate its generalization when display scale is reduced. Consequently, methods for *line simplification* based on *vertex elimination* are the main focus of the work reported here. In addition, however, QTM may be used as a way to assess the positional accuracy and intrinsic detail of vector map data, and based on such analyses to process the data in order to meet such specifications, whether derived or predefined. This notion was introduced in section 3.3, and is discussed further below. Lastly, the partitions of geographic space that QTM defines can be used to help identify congested or overlapping map features via indexing their components by location. Doing this can reduce the amount of effort needed to detect crowding and assist generalization operators that perform aggregation, collapse and displacement to rectify such conditions. The section concludes with a discussion of computational strategies for handling such analyses.

4.2.1 Evaluating and Conditioning Locational Data

In the absence of positional metadata — or even when they are available — encoding map feature coordinates as QTM identifiers can help to identify their intrinsic resolution, hence the map scales at which they can be appropriately displayed. This capability has been described and demonstrated in section 3.3, where it was shown that the Swiss canton boundary dataset used for testing seems to have on the order of 100 m resolution, such that points within it can be described with 16 or 17 quaternary digits. Given the 0.4 mm symbol separability assumption used throughout this study, this implies that this particular dataset should be displayed at scales no larger than 1:150,000. Naturally, the type of map display and its purpose and intended audience can influence this limit; the effectiveness of many base maps used in thematic cartography may not be greatly compromised simply because boundaries appear coarsely-drawn. But as a general rule, and particularly when map data from several sources must be integrated, it is quite helpful to know the range of scales over which display and analysis can usefully be conducted.

In the analyses presented in chapter 5, U.S. nautical chart data purported to be digitized at scales ranging from 1:20,000 to 1:80,000 were similarly analyzed, and found to have an intrinsic resolution exceeding 50 m, implying that it is reasonable to display them at 1:80,000 (QTM level 18) or less. The way in which this was determined is as follows:

1. Coordinates of coastlines were QTM-encoded to 20 levels of detail (ca. 10 m resolution).

2. Starting at level 20 and going down to level 10, the QTM IDs of adjacent vertices were compared, and duplicates eliminated until all contiguous vertices had unique IDs.

3. The number of remaining points were tabulated, and the QTM level at which at least 98 % of vertices were retained was selected as indicative of the coastlines' resolution.

The 98% figure is arbitrary, but it is a reasonable a cut-off, because when the level 18 QTM IDs were decoded back to geographic coordinates (using only that amount of precision to regenerate point locations at QTM quadrant centroids) no visible difference could be observed between the reconstituted and the original data at any scale smaller than about 1:25,000. Appendix figures D-1.1 and D-2.1 illustrate the results of this data conditioning procedure, although at too small a scale to show more than a few very small discrepancies between the data as input and as conditioned at QTM level 18.

Figure 4.1 displays the same parameters as figure 3.11 but for the island shorelines used as test data, shown in figure 5.1. Its x-axis measures the spatial resolution of sampling (in meters) and the associated QTM levels of detail; the y-axis measures the number of points retained at each level of detail by the quadrant culling operation described above.

Both ends of these distributions exhibit shoulders, representing effective limits to minimal and maximal simplification, and a relatively uniform slope in the central region, roughly from QTM levels 13 through 17 (1,000 to 75 m resolution). Had there been higher-frequency (larger scale) detail in the source data, the right-hand shoulder would be shorter, or move rightwards, and additional QTM digits would have been required to capture the information.

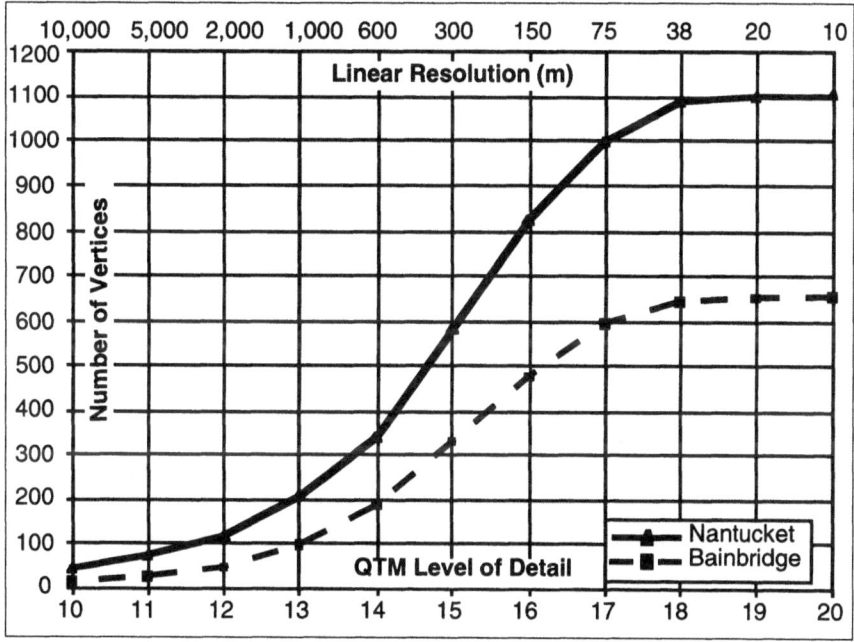

Fig. 4.1: Bainbridge WA and Nantucket MA; Number of Retained Vertices by QTM Resolution

4.2.2 Line Simplification

In the procedure for culling duplicate and closely-spaced points itemized above, no geographic coordinates are directly used or need be computed; all operations are performed on QTM identifiers themselves. In doing so, the number of vertices defining features is normally reduced, although not by very much in these instances. Wherever this happens, the effect is to regularize distances between successive vertices along the contour of a feature by eliminating some (but not necessarily all) closely-spaced ones. The effect is similar to that of a low-pass filter applied to a linear, two-dimensional distribution of data points, in that shorter-wavelength details are eliminated. Another way to think of this process is in terms of enforcing limits of visual acuity; a fixed (triangular) sampling grid of a certain density limits detail to the amount that can be perceived at a given "viewing distance" (i.e., scale). Accounts of this "natural principle" applied to map generalization in both raster and vector domains were published by Li and Openshaw (1992, 1993), but other implementations (especially raster-based ones) no doubt preceded that work, given the obviousness of the analogy to foveal vision.[1]

---

[1] A mapping toolkit developed ca. 1988 by TerraLogics (Nashua NH, USA) included a resolution-driven generalization method similar to that described by Li and

The same basic approach may be extended to eliminate vertices in order to generalize linear data at any reasonable level of detail below the one that characterizes the source data's resolution. What is "reasonable" in this context is primarily dependent on shape characteristics of features themselves. Where features are small and arcs are short, containing relatively few coordinate points, a useful limit of reduction may be reached sooner than when features span greater distances and have more complexity. In most cases, data in the latter category tends to require more generalization and to respond better to simplification efforts anyway.

An extension of this method of generalization is to apply it using QTM attractors instead of facets. Nothing else changes except for the sampling grid, which consists of hexagons and triangles instead of just triangles. The triangular attractors are identical to QTM facets at each level of detail (they are all facets whose identifiers end with zero). The hexagons, being larger and more nearly circular, agglomerate more vertices and tend to lessen sampling artifacts when features have complicated shapes. As a result, generalization using attractors almost always yields a greater amount of simplification at a given level of detail. Figure 4.2 illustrates the logic of generalizing via both facets and attractor occupancy, for fictitious data.

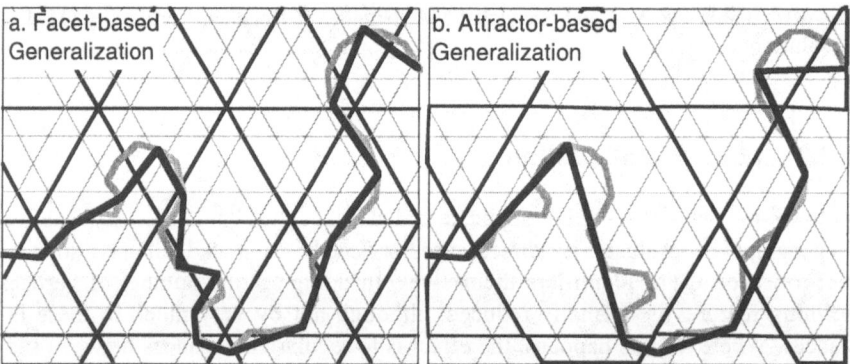

Fig. 4.2: QTM generalization via facets (a) and attractors (b); filtering elements are shown in black. QTM levels of facets (light gray lines) are indicated by line weight.

In figure 4.2, a feature (gray polyline) has been simplified by selecting among vertices that occupy a single facet (left) or attractor (right) to yield the (black) ver-

Openshaw, details of which were never published. This algorithm may still be found in mapping applications that use the toolkit, including Atlas Pro GIS (Atlas Graphics, Santa Clara CA, USA, now owned by ESRI, Redlands CA). None of these packages provide direct user control over its operation, generalization being entirely based on zoom levels. One could argue that users do not need control, as the algorithm operates below the threshold of perceptibility, assuming that display pixel density is properly specified.

sion of the polyline. This is done by identifying "runs" of contiguous vertices passing through each mesh element (facet or attractor) and selecting the most central vertex of each run as its representative point. We call this the *median vertex* algorithm, specifications for and discussions of which may be found in appendix A. Additional controls and refinements to this algorithm are possible, and are discussed in that appendix as well as in section 4.3 below. It should be noted that more than one run of vertices may occupy a mesh element, as lines may exit and re-enter facets and attractors. Each such entry is considered as the start of a new run, and processed independently of any other run, although our implementation allows one to override this constraint to a limited degree. Despite this and other refinements to it, the median vertex method is a rather generic, easily-understood spatial filter.

4.2.3   Spatial Conflict Identification

Another way in which QTM can assist map generalization — and potentially its greatest contribution — is to facilitate identification of *spatial conflicts*, or regions where map features crowd together and compete for space when map scale is decreased. QTM shares this capability with raster and quadtree-based methods, but these are rarely applied to vector data. While it is possible for a single feature to interfere with itself (and this is a major reason for applying simplification operators) most of the recent work in the field has been focused on so-called *contextual* or *holistic* generalization, as discussed in section 1.4; this research area looks at how map symbols interplay, how generalizing one feature affects others nearby. The best-known contextual cartographic problem may be *text label placement*, in which text identifying features must be located appropriately, near enough to create a visual connection but not interfering with the feature being labelled nor any others. This is naturally, almost effortlessly accomplished by human mapmakers, but requires considerable programming ingenuity for software to do well. Still, a variety of effective algorithms now exist that perform it, all of which require the close support of data structures that identify connected neighbors and/or provide access to features by location.

A quaternary mesh is useful for such purposes because it partitions space into discrete units at different scales, which can be searched efficiently and related to feature data that occupies them. It thus provides space-primary access to object-primary data, that is spatial indexing. This topic was discussed in section 3.2, followed by an exposition of QTM attractors in section 3.3, where it was shown how attractors knit together adjacent QTM quadrants that are not siblings in the hierarchy of facets. A convention for naming attractors was also shown; using the QTM identifier of quadrants that constitute triangular attractors (*tri-attrs*), or choosing the lowest QTM ID of the six members of hexagonal attractors (*hex-attrs*). This approach assigns identifiers to facet centroids and vertices, respectively. Other naming conventions, including an earlier one proposed by Dutton (1990), are possible, but will not be explored here.

Computing and comparing attractor identifiers and their contents is an option when working with QTM-encoded vector data. They need not be used, or be made a permanent part of a database. Since the sizes of attractors to employ may be freely selected by an application, there is little point to pre-defining them as data. Therefore, even though attractors define very particular regions on the earth's surface, they are abstractions that come and go as an application demands, ephemeral structures that are maintained only while they have an operational role and then dismissed.

Selecting Attractors to Evaluate   How might attractors be used to identify which locations could possibly conflict at particular scales? A naive approach would be *to compute an attractor for each vertex along every feature that has the potential to occupy the same map space* (based, for instance, on whether bounding boxes intersect in latitude and longitude). This could be done by computing attractors at the target scale to determine which ones are occupied by more than one vertex. There are several problems associated with this approach:

1. If the target scale is close to the source scale, most attractors will contain only one point, and waste memory and time to construct and evaluate.

2. If the target scale is very much smaller than the source scale, attractors will be large and will contain too many indexed vertices; there are easier ways to determine gross levels of interference.

3. Points in different attractors may still be close enough to cause interference, requiring a way to search for neighbors of attractors (of which there are either three edge plus three vertex neighbors, or six edge plus six vertex neighbors.

A refinement to this strategy would be to evaluate attractors several QTM levels below the target scale. This will generate fewer of them, each of which is more likely to contain more than one vertex. As item (2) above indicates, using too coarse a resolution will cause many "false positive" results — vertices that really are not in conflict at the target scale, but are reported as being associated at the test resolution.

It should be possible to select an appropriate resolution at which to begin evaluating attractors. One way might be to scan the QTM codes of a given primitive (or set of them) to identify the common base code, if any, that they all share (n leading digits). In the worst case no leading digits (even in the same octant) will be found in common — this happens when features straddle the 45th parallel, for example. But usually, there are a fair number of shared digits, especially when features and regions of interest are relatively small. When there are few shared initial digits, there will usually exist a relatively small set of cousins that spawn clusters of QTM IDs, within which additional leading digits are shared. The common digits in fact identify a "minimum bounding triangle," or *mbt*, that all the primitive's vertices are guaranteed to lie within.[2] During generalization, when scale change goes to the point where a polyline's *mbt* is the filtering element, such an *arc* will be swallowed whole, replaced by single points or a straight line segment. Even then, other arcs would also run through the *mbt*; these need to be examined for conflicts at a test resolution.

At its first sub-level, an *mbt* contains 1/6 of three hex attractors and one tri attractor, but this is probably too coarse-grained to start with. By dissecting the *mbt* again, 10 attractors come into play: 1/2 of 3 hex attractors, 1/6 of 3 hex attractors and 4 tri-attractors (this amounts to 16 triangular attractors, the number of facets in two subdivisions; see figure 4.3a). The process could begin at this point, keeping track of the contents of 10 attractors contained by (or intersecting) the *mbt*. This method can then be applied recursively to handle included facets(treating each the same way as the *mbt*) in a divide-and-conquer strategy, stitching results together by associating neighboring attractors.

We can begin by evaluating the four tri-attractors, as they are the most centrally-located and are likely to contain at least some of the vertices. We know what they are called, as well (their AIDs are their QTM IDs at respective levels) without having to compute much: if the QTM ID of the bounding triangle is Q, the triattrs two levels down are Q00, Q10, Q20 and Q30. We also know what the QTM IDs of their neighbors are (Q01, Q02, Q03 ... Q33), and which of these are neighbors to each other (they must share the same last digit). And since we can directly evaluate the AID of any of these neighbors, we can thus name each of the ten attractors in the region.[3] Figures 4.3 and 4.4 show the shape, arrangement and constituent quadrants of these 10-attractor neighborhoods.

An alternative way of deciding what level of attractors to use, and one that is more globally consistent (i.e., is not based on each feature's extent), would be to consider symbology at target scale. One might determin, for example, the width (in ground units, e.g. m) of the largest line weight to be displayed, and build attractors at the QTM level having that linear resolution. A somewhat coarser resolution (1 or 2 levels lower) could be used to be more confident of detecting conflicts, and then refined as needed.

---

2 When features cross high-level QTM quadrants, causing their *mbts* to be much larger than their physical extents, it is usually possible to allocate two or more smaller bounding triangles that contain more-or-less contiguous portions of them, and evaluate these quadrants individually as pseudo-*mbts*.

3 There are two ways to compute AIDs: 1) evaluate the QTM ID of the appropriate vertex of the facet being examined (see discussion of the *qtmEval* algorithm in appendix A); 2) use lexical substitution rules to identify the five node-incident neighbors of the given facet that share its last digit, selecting the one with the smallest ID. Results of the two methods are equivalent, but the former requires some floating-point computations and computes an AID only, while the latter is based on string manipulation, and yields the QTM IDs of all member facets as well as an AID.

a. A 12-point arc and its minimum bounding triangle (mbt)	b. The 12 vertices are found to occupy 6 attractors
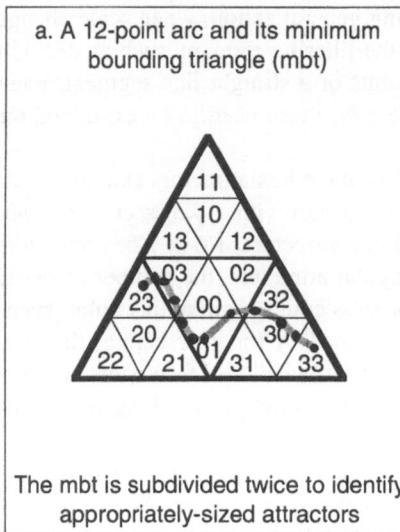	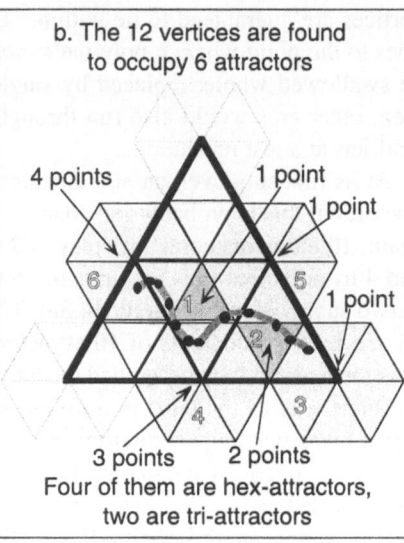
The mbt is subdivided twice to identify appropriately-sized attractors	Four of them are hex-attractors, two are tri-attractors

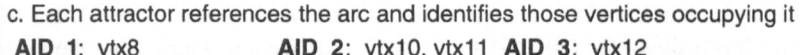

c. Each attractor references the arc and identifies those vertices occupying it
AID_1: vtx8   **AID_2**: vtx10, vtx11 **AID_3**: vtx12
AID_4: vtx5, vtx6, vtx7   **AID_5**: vtx9   **AID_6**: vtx1, vtx2, vtx3, vtx4

Fig. 4.3: Associating attractors to an arc using a dissected minimum bounding triangle

Specifying the Contents of Attractors   Having named an attractor occupied by a vertex of a feature, this information needs to be put somewhere (at least temporarily) in order to itemize the vertices that "belong" to a given attractor. As more features are processed and additional vertices fall into an already-identified attractor, this collection will grow — not without limit, but arbitrarily large. It grows in direct proportion to data density, both in terms of feature primitives and of points belonging to them. It also will be larger for bigger (lower-level) attractors than for smaller ones. There are two design decisions that this raises:

1. What is the minimum set of data that need to be stored when a vertex is added to an attractor?

2. How can these sets of observations be organized for efficient search and retrieval within an application?

The second question — a data structure design problem — will be addressed in the following section. The first one can be answered by considering the role of attractors in the generalization process. As the purpose of using attractors is to identify features, primitives and vertices that are competing for map space, we need to be able to trace back to these objects from any attractor they occupy. The minimum data to accomplish this would consist of:

1. Primitive (point, arc, polygon) ID

2. Vertex position within the primitive

It may also be of interest to know the QTM IDs of attractor occupants, as well as the ID of the feature to which the primitive belongs. However, the two items above are sufficient to derive this information given the data model being used; once the primitive is known, the ID of the given vertex is easy to find, and all features that share the primitive can be identified as well. The particular feature involved (i.e., the one whose processing caused the data to appear in the attractor) will not be identified, however, should the primitive be shared by several of them. It really does not appear to matter which feature caused the attractor to be occupied, because if the primitive must be modified to generalize it, all features that use it will undergo the same change.[4] Nevertheless, resolving conflicts at attractors may depend on the properties of features, in particular on their feature class and its attributes.

When one needs to query the QTM IDs of vertices occupying an attractor, it should be sufficient to index into the primitive to find this data, without keeping a copy of it in its attractor(s). Therefore, two data items (primitive ID and vertex number) are all that must be stored, although for any given attractor there can be any number of these tuples. The first item (Primitive ID) can either be represented as a memory address where the primitive's structure is allocated, or as an internal identifier that is assigned when the primitive was read in. The second item (vertex index) is simply a positive integer, which can be a 2-byte quantity if primitives are limited to 64K vertices each. Figure 4.3 (in which attractors are identified by index numbers rather than by their true AIDs) illustrates mapping vertices of a primitive to attractors.

Attractor Data Management Under the above protocol, each vertex that occupies an attractor can be identified using six to eight bytes of data. The data can be structured as a linked list; in our testbed, all such lists consist of tuples of the form:

List_element ::= [object_pointer][object_identifier][list_pointer]

The tuples comprise a singly-linked list, which is sufficient to reference features and primitives. It entails an overhead of four bytes per list element, to store the pointer to the next one. If there are more than a few dozen items, using such lists might degrade performance when searching for specific elements.

---

[4] The data model being used allows n:m relationships between primitives and features; in it, primitive objects contain identifiers of all feature objects which use them, allowing the tracing of generalization effects through any number of related features. The data model is described in section 4.4.1.

De-referencing attractors is a more demanding task, because every time a vertex is added to an attractor, the application must see if the attractor is already occupied. Consequently, attractors must be easy to find once they have been allocated, and easily modified, able to grow as large as necessary to store all the vertices that may collect there. Some standard ways of doing this include:

1. Simple linked lists;
2. Tree structures
3. Hash tables
4. Directories

Assuming that the key to storing or retrieving an attractor (and its data lists) should be its AID itself, which is unique and descriptive of location, we consider each of the above alternatives in turn.

Linked Lists. Using lists would make managing attractors similar to managing other types of data in the testbed, resulting in the least code overhead. It would provide a simple way to itemize all occupied attractors, and to delete attractors that become empty or are otherwise no longer of interest. But as the list of attractors can grow large, and as $O(n)$ time is needed to locate a particular item in a list, this is not an efficient solution for searching, even if the list is ordered by AID. One way to deal with the access problem is to maintain a set of AID lists, a different one for each level in the QTM hierarchy. Regardless of how many lists are kept (which would be limited to 29 levels of detail), each list item would contain an AID, a pointer to the head of a list of vertices, and a pointer to the next attractor list element, requiring $8 + 4 + 4 = 16$ bytes per AID. Still, any given list could still grow large, as many evaluation strategies might only refer to a single level of attractors.

Tree structures. Organizing attractors into trees would make it faster to locate particular AIDs, and might facilitate access to their parents, their children and their neighbors. As was described in section 3.3, the inheritance patterns of attractors are complex, and relatively fewer attractors have a single parent as the hierarchy deepens. The fact that most attractors have no simple lineage may not be a problem in itself; if all that tree storage is intended to facilitate is access by AID, their provenance may not matter. The major computational concern is the need to keep the AID tree(s) balanced, lest they devolve into linear lists, and lengthen the time needed to insert, delete and search for AIDs.

Hash Tables. Hashing — computing a constrained, pseudo-random address from a record's identifying key — avoids the overhead of allocating memory in small units by reserving a block of memory and incrementally filling it with keys until it is exhausted (at which point new memory can be allocated, and the table re-hashed). It can also be efficiently implemented in a paging environment, where it is assumed that portions of a hash table reside on secondary storage. To avoid collisions, a hashing function is needed which will map any given set of AIDs to a relatively uniform distribution of hash keys. While searching a tree for a specific AID normally takes $O(\log n)$ time, searching a hash table takes constant time (or possibly greater, depending on how often collisions occur and how they are resolved). As is the case for tree structures, many designs for hash tables have been invented: some are suited for RAM access, others for disk access; some have good performance in the event of deletions, others do not; some only perform well when the universe and distribution of access keys is known beforehand, others make few assumptions about these matters, but may suffer more collisions as a result.

Directories. A directory is a collection of a finite number of names of objects of the same type, usually ordered according to some principle (such as lexical order, date of creation or size), and includes a pointer to each object's storage location. In this respect, a digital directory is just like a directory of businesses in an office building, which are usually ordered alphabetically and provide a floor and suite number as an address for each entry. Directories are most useful where the stored objects have a common naming convention and persist for a relatively long period of time, such as files. There are a number of implementations possible, including linked lists, arrays, linked arrays and extendible hash tables. At first glance, it might not appear that attractors are likely candidates for directory storage, given how uncontrolled their growth could be, but if they are evaluated according to appropriate protocols, their numbers can be contained.

Earlier in this section, it was suggested choosing attractors to be evaluated starting with groups of ten of them occupying a bounding triangle defined by one or more QTM-encoded primitives. This will identify all attractors that are actually occupied at that level. Particularly crowded attractors can be decomposed to resolutions at which decisions can be made about individual vertices, which presumably would let their contents be filtered to the target scale. Therefore, the directory could be hierarchical, not in the sense that smaller attractors are descended from particular larger ones, but in regard to the decomposition of the QTM grid, where inheritance is strict and always 4-to-1. But keeping the workload within bounds — and therefore limiting the proliferation of directories — is not a simple challenge.

Figure 4.4 illustrates a possible structure for such directories. In 4.4b, the first column holds abbreviated AIDs, "N" specifies how many primitives are indexed to that AID, "&data" points to a linked list holding the primitive references, and "&next" if not nil, points to a child directory block spawned by the given block element. Each of a block's 10 member attractors can generate such a child, and every new directory level represents a fourfold linear decomposition of an *mbt*; 16 facets and (all or portions of) 10 attractors arise from every facet shown. The next iteration (two levels higher) will have 256 facets, coded as 160 attractor entries. However, these 160 consist of 64 tri-attrs and 45 hex-attrs (109 in all), a partitioning which is not helpful. To avoid fragmenting hex-attr vertex data across several separate child directory blocks, the "&data" pointers in all entries that identify the same AID should reference a single linked list.

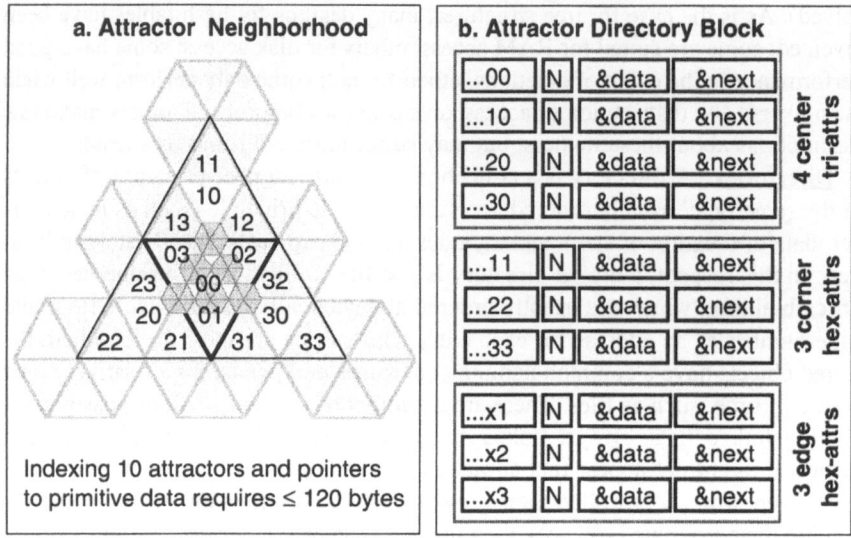

Fig. 4.4: A Directory for Associating Attractors to Primitives. The next more detailed block of attractors (two levels up) for the facet labelled 00 is indicated in the center of 4.4a.

The directory shown above is efficient for cataloging the contents of attractors, but does not provide direct access to the attractors occupied by a given primitive. In general, data structures for attractors should be based on how attractors will be used by an application. If a simple collection will suffice, why not use a list? But we need to access attractors as well as store them, and lists are not so efficient for finding things. For spatial indexing purposes, we may wish to know parents, children or attractors with the next smaller or larger key. If so, it might be useful to construct a tree based on access keys. If we are interested in working with attractors in pre-defined neighborhoods or across a limited set of objects, a directory-based approach may be most efficient. And if the major concern is to identify the data belonging to a specific or arbitrary attractor, we might better use a hash table.

Implemented Approach. The prototype software created for this project uses attractors only for generalizing individual arcs and for indexing nodes (endpoints of arcs), to assist spatial indexing and to determine which arcs chain together or are simple polygons. The test data used consists of unstructured polygons and arcs ("spaghetti") into which continuous coastlines were digitized on a piecemeal basis. Near-source-resolution attractors of nodes were computed and compared in order to link arcs together, eventually forming polygons. Generalization was then performed on the polygons, to minimize the number of "sacred" vertices (because endpoints of arcs are always retained when generalizing). This prevented test results from being biased by arbitrary segmentation of the coastline data.

For the purpose of identifying polygons and linking arcs, it is only necessary to compute AIDs for endpoints, not for every vertex. This keeps the directory to an easily-managed size. However, the directory scheme shown in figure 4.4 provides indexing from AID to primitive IDs, but not back again. Arc linking requires both, and for this purpose a compound directory — not organized by *mbts* — was designed, which is illustrated in figure 4.5. Were we to also employ this structure for detecting spatial conflicts, dissected *mbts* could still be used to select attractors to be evaluated.

The implemented attractor/primitive directory differs somewhat from the structure shown above. First, *mbts* — and hierarchies of attractors within them — are not modeled, nor are attractors organized in groups of ten. Instead, the directory has a linear, bipartite structure that is allowed to expand as much as necessary. The first part indexes primitives by AID; the second part indexes attractors by primitive identifiers (PIDs). Each one has an index for the primary key that is kept in sorted order as data is inserted and deleted. The index array points to "slots" where data for the key is kept. Among other things, each slot stores the address of a list; when the key is AID, the list contains the PIDs and vertex indexes of all primitives that were found to occupy the given attractor. When the primary key is a PID, its directory data consists of a count and list of AID slots that the primitive's vertices have been found to occupy.

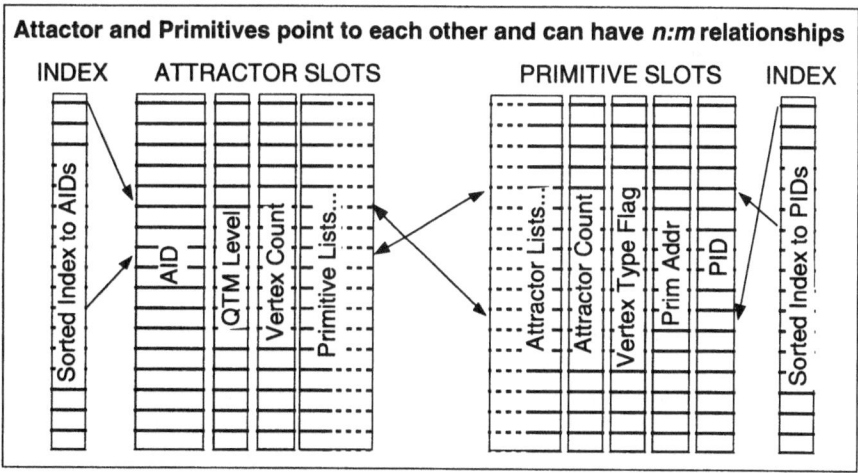

Fig. 4.5:   Structure of Dual Attractor/Primitive Directory Used in the Study

The sorted indices speed access to identifiers by enabling binary search. Larger attractors have shorter AIDs, so that attractors are ordered by size, when it differs. In figure 4.5, the end nodes of a primitive are indicated by arrows on the right, and the attractors these nodes by arrows on the left. Arrows in the middle show how AIDs and the PIDs point to one another's directory slots. Any number of list elements (within memory limitations) can be stored on either side, each pointing to a corresponding entry on the other side. In addition, the PID side of the directory contains a flag for each entry to signal if the primitive has attractor data for (1) nodes, (2) its centroid, and/or (3) individual shape points, to help quickly assess the role of attractors stored for each primitive. Not shown in figure 4.5 are certain housekeeping details, such as fields to store directory statistics and caches for recent hits.

Using this structure, it is possible to record all attractors (of any level) that a primitive might belong to, and each one will also be indexed back to all the locations on the primitive where it was evaluated. Of course, doing so for every vertex in a set of features would be cumbersome, possibly straining memory resources; that is why starting several QTM levels down from a primitive's *mbt* might be a useful heuristic when searching for spatial conflicts.

4.3   Parameterizing Line Generalization

Most line generalization procedures have a parameter that controls the degree of generalization (e.g., a scale factor, a bandwidth tolerance or a point reduction ratio), and may have several others that also inform the process. For example, a culling distance (to enforce a minimum spacing between points), a minimum area criterion (to drop out small polygons) or an angular tolerance (for example, to preserve the shapes of right angled objects) may be found in the list of calling parameters. In general, the more parameters an algorithm provides to a user, the more control he or she has over the results. The price that users pay for this may include a degree of perplexity over how to set each parameter, and possible uncertainty about how strongly parameters may interact. But no matter what algorithm is used and how many parameters it has, some results (including incorrect ones) will always be beyond the user's control.

4.3.1   Regulating the Amount of Detail

The widely-used Douglas-Peucker (1973) line simplification algorithm — actually first described by Ramer (1972), and referred to herein by both names (Ramer-Douglas-Peucker, or *RDP*) — has been widely studied and implemented in many GIS environments. Part of its popularity is due to its ease of implementation, part to its often intuitively-sensible results (White 1985, Marino 1979), and part to the fact that it requires only one control parameter, the width of a band around trend lines (also referred to as *anchor lines*). However, although this tolerance may be given in either ground or map distance units, users may have a difficult time conceptualizing how it will affect point selection, and often may have to attempt a number of trials and display the results in order to achieve the level of detail they desire.

The RDP algorithm has been adapted to enable any degree of point reduction desired. Van der Bos and Van Osteroom (1989), Van Osteroom (1993) and Cromley (1991) have done so to make line simplification faster and more pre-dictable by hierarchically ordering the vertices of arcs. This is done by processing an arc with the RDP algorithm with a zero tolerance and storing the distance from each point to its local trend line when the point is selected. The distances are then ranked to assign a level of importance to each point. When retrieving the arc, it can be generalized from N to n points by selecting vertices with importance in-dices from 1 to n, saving time and achieving precise control over global point density, but not necessarily over local point densities. The same strategy is possi-ble with QTM-based methods, as will be discussed in section 4.4.

4.3.2   Controlling the Quality of Detail

Automated line generalization does not often succeed very well when the ratio of scale change is high, say greater than 1:8. This makes it difficult to construct, for example, a 1:250,000 display of data originating from a 1:25,000 topographic source map. This has often been mentioned in studies of RDP, but extreme scale change is not easy for any algorithm to handle. The general problem seems to be at least twofold: (1) unintended intersections of lines with themselves and other objects are more likely to occur when they are simplified to a high degree using a non-topological approach (all methods other than *Whirlpool* (Chrisman et al 1992) and that of de Berg *et al* (1995) ignore topology); (2) in the case of RDP and simi-lar algorithms — including "sleeve-fitting" methods such as (DeVeau 1985) and (Zhao and Saalfeld 1997) — a large tolerance can "swallow" large amounts of line detail along relatively straight sections of features, where no intermediate point is deemed significant enough to retain. As a result, large extents of gentle, sinuous lines can be obliterated, creating an unnatural appearance.

The QTM-based line simplification method described below cannot guarantee topological consistency, although this problem seems to occur relatively infre-quently, usually where shapes are contorted. It does, however, maintain a rather uniform degree of detail along lines, avoiding the second kind of problem itemized above. The workings of this method is described in the next section, with further details provided in appendix A.

4.4   Vector Line Generalization Using QTM

As described in sections 4.1.1 and 4.1.2, a quaternary mesh is used as a spatial filter to generalize vector data that pass through it. Several algorithms that do this have been developed and tested; some of these make use of "enriched" and "enhanced" QTM IDs (described in section 3.1.4) to help decide which vertices to select and delete when changing scale. For single-theme maps, attractor- and quadrant-driven simplification are able to produce quite reasonable generalizations, but obtaining the best possible results requires some understanding of methods as well as trial-and-error. This should be possible to automate as our understanding grows.

4.4.1   Data Modeling

The QTM generalization methods described here are supported by a general, feature-based data schema (see figure 4.6), chosen for its flexibility in modeling map features, even those containing features of mixed dimensionality. It can also handle information provided about topological relationships between and within map features, but does not require it[5]. Features may be designated as arbitrary collections of primitive elements (points, arcs, polygons); an element may play a role in more than one feature (be a so-called "shared primitive") as well as being topologically linked to other primitives, hence features. Although topology and sharing can be exploited for generalization purposes — both indicate what other elements may be affected by alteration of a given one — our experiments did not explore these techniques.

Feature classes, such as transportation, hydrography and land use, as well as their sub-categories, may also be modeled, and provide a higher level of abstraction to the data schema. In our model they are called *feature sets*, and may be defined hierarchically. Like features, which can share primitives with one another, feature sets may also have overlapping contents; the same feature can belong to more than one set. In our experiments, in which all geometric data represented shorelines, and no attribute information was available, three pseudo-feature-classes were defined: *arcs*, *linked arcs* and *polygons*. They can be considered as proxies for more specific class distinctions that a full implementation would employ.

As implemented, nearly everything in this data schema is a *list*, either of features, primitives, vertices or attractors. All lists are either ordered, fixed and linear, such as used for vertices along a primitive, or are unordered, extensible and singly-linked, such as used to enumerate primitives that comprise a feature, or features making up a feature set. The former list type is equivalent to an array. The format of the latter type of list is described in the discussion of Attractor Data Management in section 4.2.3.

The basic organization of this data model is presented in figure 4.6. Several of its aspects require comment. First, at the top is the data *Catalog*, a central registry for all primitives, features and feature sets in the current dataset. It is through the

---

[5] This is done by providing each primitive element with a minimal set of fields: *from*- and *to*- nodes, and *left*- and *right*- polygon IDs (point sets have a field that identifies a containing polygon instead). Feature objects do not have these descriptors, as all topology is modeled as a property of primitives.

catalog that all these entities are accessed and their existence confirmed. When primitives are deleted and features or feature sets become empty, it is through the catalog that the integrity of relations is enforced.

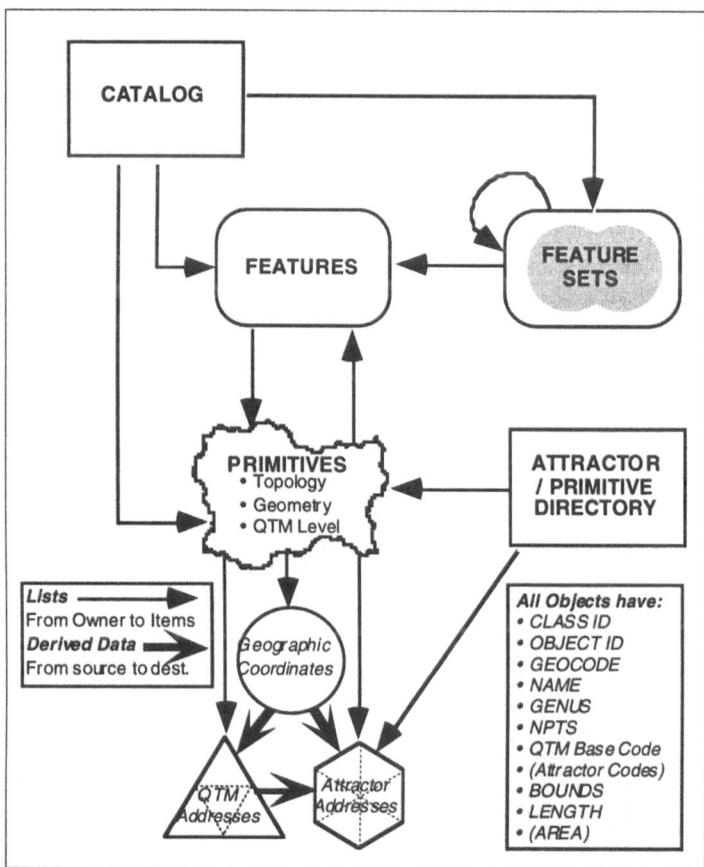

Fig. 4.6: A Feature -based Spatial Data Model for Map Generalization

At the right of figure 4.6, the *Attractor/Primitive Directory* (described in the prior section) is shown connecting attractors and primitives. Some attractors exist only in this directory, others exist as lists, in parallel to QTM IDs and with a 1:1 correspondence with QTM-encoded vertices. But attractors are scale-specific, and may need to be regenerated at different precision. Because their specifications are subject to change, vertex attractor lists (the hexagon near the bottom of figure 4.6) are created on-the-fly and are usually deleted after being used for generalization.

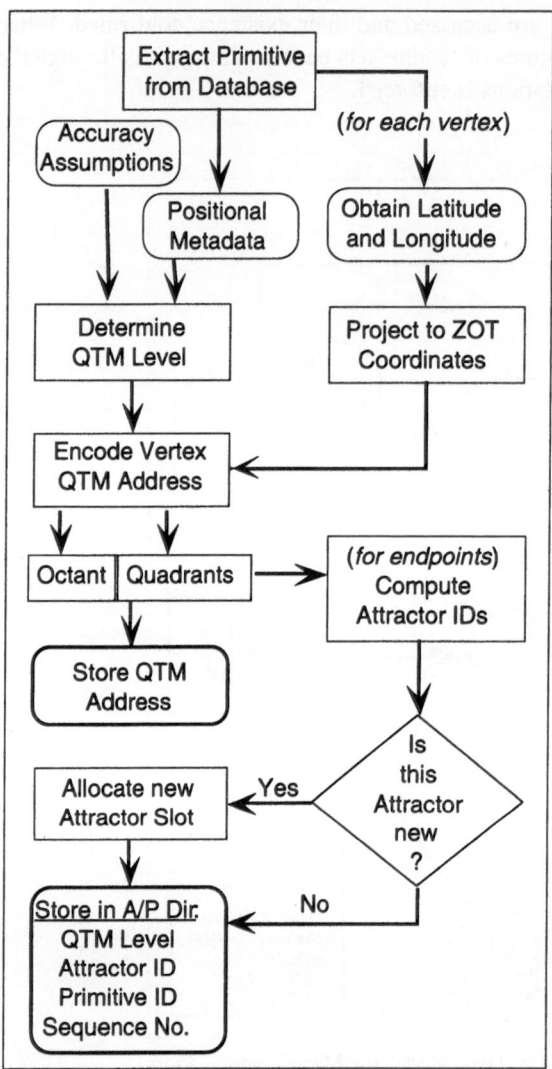

Fig. 4.7: Computing QTM Hierarchical Coordinates for vertices of a primitive

The flow of control for entering map data into this environment is illustrated in figure 4.7. Each primitive is processed separately; enough memory is allocated for QTM IDs to store all its vertices. Once all vertices are encoded, the address of this list is copied to the primitive, which in effect owns these codes. At this point its geographic coordinate data can be deleted from memory, as it can be regenerated as needed from the list of QTM IDs. The primitive also has a slot in which to store the address of a list of vertex attractors, that could be computed in parallel with the QTM IDs. This is not normally done, as the appropriate attractor size needs to be determined by the application, and is related to the task and target map scale, which may not be known initially.

Several sub-processes are omitted from figure 4.7. The first one, described in section 4.2.1, involves scanning the list of QTM IDs at its full resolution to identify duplicate codes for successive vertices, and eliminate all but one of the duplicates. The second one — which is optional — is to compute a measure of each vertex's local geometric importance (see section 4.4.3). This requires a geometric analysis of the primitive's coordinates (not QTM codes), and results in a categorization of vertices according to their degree of sinuosity. Once computed, each vertex's sinuosity measure is stored in a byte-field within its QTM ID.

4.4.2 Parametric Control of QTM Generalization

Unlike the one or two parameters that control most generalization algorithms, eight independent parameters work together to control vertex precision, selection and placement when simplifying QTM-encoded features. These are itemized below:

1. QTM Encoding Depth
>   *Controls maximum coordinate accuracy available*
>   — From 1 to 29 doublings of resolution
>   — Too few levels truncate essential precision
>   — Too many levels encode spurious precision,
>   but may not be harmful

2. QTM Retrieval Depth
>   *Principal control over point resampling density for display*
>   — Less than or equal to QTM Encoding Depth
>   — Specifies linear resolution and output scale
>   — Primary determinant of output point density

3. QTM Decoding Depth
>   *Controls accuracy of selected points*
>   — Less than or equal to Encoding Depth,
>   greater than or equal to Retrieval Depth
>   — When low, moves points to coarser grid locations
>   — When high, points are not visibly displaced

4. Number of Vertices to Retain per Mesh Element
>   *Controls degree of simplification, along with Retrieval Depth*
>   — Specifies maximum number of points to keep from each run
>   — Vertices chosen up to this limit by regular sampling
>   — Within limit, number to keep is decided by radical law
>   — Affects large scale changes much more than small ones

5. Hierarchical/Nonhierarchical Generalization Strategy
>   *Controls consistency of simplification across levels*
>   — Hierarchical strategy enforces top-down consistency,
>   eliminates more points and takes more time
>   — Nonhierarchical strategy is QTM level-specific,

more flexible and faster

— On-screen zooming should use hierarchical elimination

6. Quadrant/Attractor Look-ahead

Helps to compensate for arbitrary placement of QTM grid

— Identifies segments that exit and re-enter a mesh element

— Look-ahead can be over any number of points

— Can eliminate spikes and crevasses in linear primitives

— Tends to increase the degree of simplification

7. Vertex Placement within Facets

Controls consistency and coarseness, has specific uses

— At facet centroid (default assumption, at decoding depth)

— At a facet vertex (aliases facets; used to compute attractors)

— At weighted centroid (could be used to displace features)

— At descendent facet centroid (closer to encoded location)

— At random location in facet (not used, not recommended)

8. Vertex Hinting

Overrides point selection within quadrants or attractors

— Models importance of each point in primitive elements

— Derived from local geometry of source primitives

— Stored as optional metadata for all vertices

— Used to select points, or may be ignored when generalizing

Some of these parameters are more constrained than others, and certain ones constrain others. Retrieval Depth (2) constrains decoding depth coarseness (3), and both are constrained by QTM encoding depth (1). Also, if vertex hinting (8) is used to select retained points, this overrides the default resampling method, although it does not change the number of points retained. Although the range over which parameters vary may not be large, changing any one of them can have unintended effects, even if its use is understood and its value is well-chosen. Understanding how the parameters can interact is not simple, and at this point requires some direct experience with this approach to feature simplification.

Implementation. The algorithm that conducts QTM simplification has been named *weedQids*, which is specified and further discussed in appendix A. This algorithm represents a distillation of lessons learned from various experimental prototypes, and retains the essential elements of these explorations, controlled by some of the parameters described above. It is the core of a QTM generalization engine that has other, peripheral parts. *weedQids* takes a list of QTM IDs specifying the vertices of a primitive as input, performs point elimination/selection, and encodes its decisions in the QTM identifiers themselves. Other portions of the engine then process the IDs to analyze results and output geographic coordinates for all points retained by *weedQids*.

As appendix A describes, *weedQids* is called with parameters 2, 4, 6 and 8 above plus other data, and returns generalization results in QTM IDs along with the number of points eliminated from each primitive. Along with the list of QTM IDs, it is given the number of vertices in the list and the QTM level to which to simplify, and is also provided with a parallel list of mesh elements to use in filtering the primitive. The mesh elements (or *mels*) are either (1) a copy of the primitive's list of QTM IDs (when filtering by QTM quadrants) or (2) a list of AIDs for the primitive's vertices at the level of detail being filtered to (for attractor filtering). In a single pass through that list, the function identifies sets of contiguous vertices that occupy a single *mel*, and once the *mel* is exited, eliminates all but one or a few vertices by setting a flag in their QTM IDs to indicate the ones that should not be retained. The following pseudocode indicates how *weedQids* is controlled by its calling function —identified here as *weedPrim* — which would be called once per primitive element to be generalized.

```
int weedPrim(const primPtr prim, const int useAttrs,
   const int level, const int ahead, const int
   hint, const int multiLev)

   int        npts, lev, lev2, zapped, retained
   int64Ptr   qids, mels

   npts = getPrimNpts(prim)
   qids = getPrimQids(prim)
   if (not useAttrs) mels = qids
   if (multiLev) lev2 = getPrimQlev(prim)
   else lev2 = level
   retained = npts
   for (lev = lev2 down to level)
       if (useAttrs) mels = getPrimAids(prim, lev)
        zapped = weedQids(qids, mels, npts, lev,
        ahead, hint)
       retained = retained - zapped
   end for
   return retained
end weedPrim
```

This algorithm does the following, invoking functions to query a primitive to access (or generate) its private data, as required:

1. Determines the number of points in the primitive

2. Extracts the address of the primitive's list of vertex QTM IDs

3. When filtering via QTM facets, identifies these as *mels*

4. Sets parameters for hierarchical/non-hierarchical elimination   (by setting lev2 ≥ level)

5. Extracts (or initiates creation of) lists of attractor *mels*, if caller requested via

```
useAttrs
```

6. Calls *weedQids* once (or more, for hierarchical selection)

7. Returns the number of the primitive's remaining vertices

Within a *mel*, *weedQids* initially selects vertices deterministically, attempting to spread retained ones as evenly as possible across it. The number to be retained is by default determined by the radical law (Topfer and Pillewizer 1966) as the square root of the number in the run. When only one vertex is retained per *mel*, the median one of each run will be selected. When more are retained, vertices are sampled regularly. This regularizes the spacing of retained points to maintain a constant resolution, but only coincidentally selects the most characteristic or geometrically important points. However, when the *hint* parameter is non-zero, *weedQids* will then re-select vertices by referring to metadata embedded in QTM IDs, overriding some or all that were chosen via systematic sampling.

Hierarchical Selection. Simplification of primitives may be performed with *weedQids* at a specific target scale (*Retrieval Depth*), or by working through all the scales between *Encoding Depth* and *Retrieval Depth*. Both approaches are useful, with the latter one providing greater degrees of simplification, as well as assuring that point selection is strictly hierarchical, as discussed in section 4.3. This is how parameter 5 above is implemented — not as an argument to *weedQids*, but to the function that calls it; for non-hierarchical selection, the deletion field in each QTM ID is reset to zero before calling *weedQids*, which then is executed once at *Retrieval Depth*. To perform hierarchical selection, the QTM ID delete fields are cleared initially, and *weedQids* is called iteratively from *Encoding Depth* down to *Retrieval Depth*, allowing deletions to accumulate. While the same vertex can be eliminated at more than one level, this process leads to a greater degree of elimination because its effects are additive.

The effects of hierarchical deletion are globally more consistent, because a vertex removed at a higher level will never reappear at a lower one. For this reason it is more suited to interactive zooming, even though it is slower to execute. The speed problem can be helped by pre-generalizing primitives, storing in the delete field of each QTM ID the lowest level at which the vertex should be displayed, rather than a simple binary flag. A field of five bits is sufficient for this purpose. When displaying primitives, a vertex will be selected if this value is equal to or less than the QTM level currently being mapped, else the point will be skipped.

4.4.3   Employing Vertex-specific Metadata

Even though mechanical point selection by regular sampling often yields quite acceptable generalization results, there are many cases where this process can be improved. A method for active — and presumably more intelligent — point selection has been developed, and is described below. Not surprisingly, it involves setting several additional parameters, for which optimal values may not be easy to determine as they can vary with respect to geometric contexts.

Parameter 8 of the list at the beginning of section 4.4.2 is called *Vertex Hinting*, as it involves using "hints" (heuristics) about the significance of primitives' vertices to steer the line simplification selection process. A hint is simply an integer attribute that describes a vertex, implemented as a byte field in a QTM ID. The attribute can model any locational quality one wishes, but the measure needs to be specific to each vertex. Our tests used a local heuristic that quantifies *sinuosity* (curvilinearity) in each vertex's neighborhood, which is defined as a fixed range of points around each given one along a primitive. The raw statistic ranges from one (where points are collinear) to infinity (space-filling curves), but it is usually less than two (characterizing the ratio of an equilateral triangle's legs to its base). It is then non-linearly transformed to range from zero to one, and scaled to an integer range — typically from one to less than ten — for storage and use in QTM IDs.

Figure 4.8 shows how the algorithm for computing sinuosity works. The procedure requires two parameters: (1) a minimum lag distance from the measured vertex (m), and (2) a maximum lag distance (n). Line lengths and anchor line lengths are computed across lag intervals and ratioed, and ratios from the different intervals are averaged. The lag distances m and n specify how many points before and after the vertex being hinted will be used in computing an average sinuosity. By keeping m and n both small (e.g., below ca. 4), the measure emphasizes conditions local to each vertex, and the resultant SVs can manifest abrupt changes in value. Larger values de-emphasize the importance of the central vertex, and resultant values of SV will be less erratic and more serially correlated, not only because the neighborhood for the operator grows longer, but because several runs are averaged.

The raw values for local sinuosity are unconstrained real numbers that need to be scaled to integers. Most of the time, values for SV will be less than about 1.2. The distributions of SV within features examined to date appear to be log-normal, highly skewed to the left. As the sinuosity statistic is used to categorize individual vertices into homogenous groups to assist in generalizing them, a classifier is needed which is non-linear. Several such classifiers are compared in figure 4.9; all are based on the statistics ($1/SV$) or ($1/SV^2$), which transforms SV into a range from zero to one.

A Measure of <u>Sinuosity</u>, *SV*, is computed for each vertex *V* along a polyline by constructing a ratio of distance ±*k* vertices along the line to the length of an anchor line centered at the given vertex:

$$SV_k = \frac{\Sigma d_{i,i+1}}{d_{v-k,v+k}}$$

where *k* is a lag parameter > 0, and *i* = *v-k* -> *v+k*-1.

SV varies from 1.0 for a set of colinear points, to ∞ in the pathological case where point *V+k* is in the same location as point *V-k*. Normally it is less than 2.

At right, *S5* and *S6* are computed for *k* = 2, at vertex 5 and vertex 6 respectively:

$$S5_2 = 1.443$$

$$S6_2 = 1.252$$

A more robust estimate of *S* is obtained if several values of *SV* are made across a small range of *k*, then averaged together. This is better than using a large value for *k* (which, too, would be more stable), as it remains more local. We notate this as:

$$SV_{m,n} = \frac{\Sigma SV_{k=m,n}}{n-m+1}$$ For *m* = 1 and *n* = 3, this averages the *S5* ratios:

$$S5_1 = 1.123$$

$$S5_{1,3} = 1.300$$

$$S5_2 = 1.443$$

$$S5_3 = 1.313$$

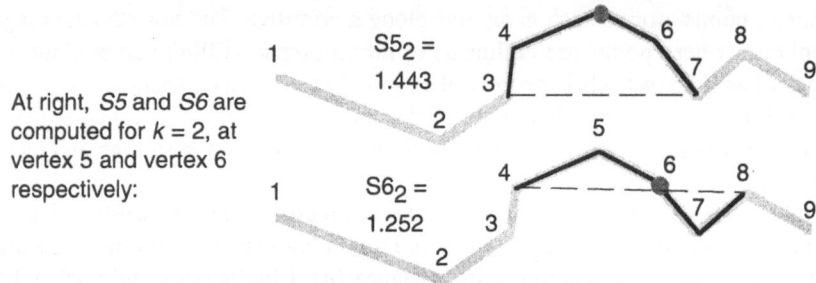

Fig. 4.8: Estimating local sinuosity at specific polyline vertices

The transform that seems to provide the greatest discriminatory power is $(1 - 1/SV^2)^{1/2}$, the most steeply-rising of the four curves shown in figure 4.9, and the one used in this study. In the above plot, the range of the transformed values has been segmented into three equal classes, labelled *low*, *medium* and *high* sinuosity. For certain purposes three classes are sufficiently descriptive while for others, more classes should be computed. We always use equal-sized partitions of the transformed value of *SV*, and normally limit the number of classes to ten. We identify these integer values as *CSV*s (classified sinuosity values).

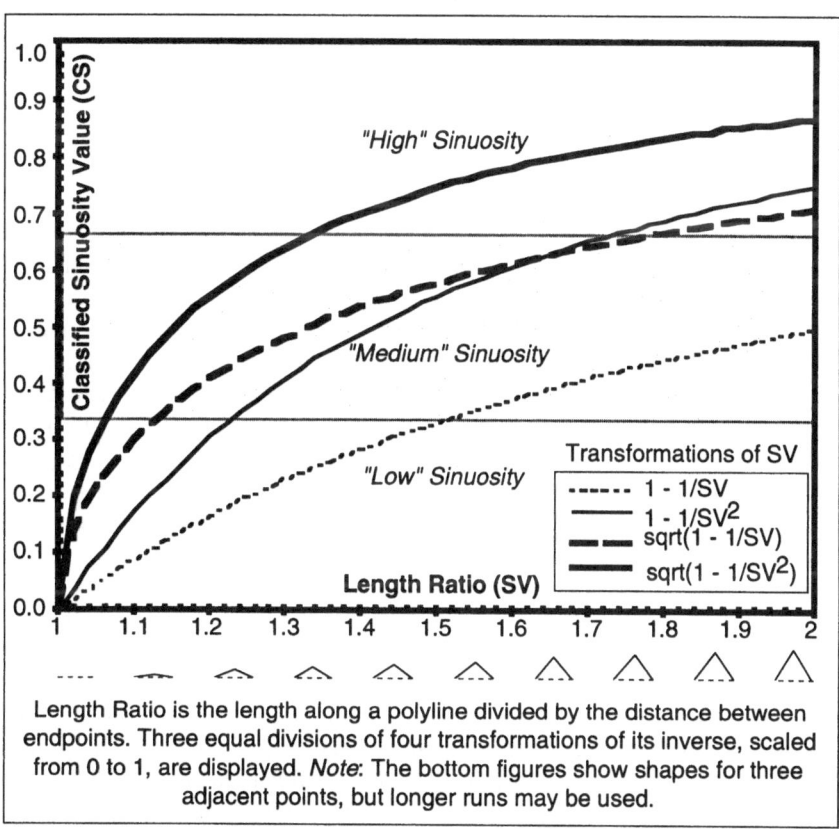

Fig. 4.9:   Non-linear transformations of sinuosity compared

Use of Sinuosity Classification. As mentioned earlier, the *CSV*s are computed for each vertex in a primitive and stored in a field within each respective QTM ID. Normally the computation of *CSV* is made at the time primitive coordinates are read in and encoded with QTM, as the geographic coordinates (via a prior transform to a plate carrée projection) are needed for the sinuosity computations. However, this can also be done later on by regenerating coordinates from QTM IDs, using any desired level of detail, point placement method, etc. The usual reason for computing sinuosities is to exploit them when selecting vertices during generalization. But there are other analytic uses for such information, as will be described below.

Point Selection. The *weedQids* algorithm uses vertex hints to override its mechanical sampling process, should the caller tell it to do so. For this purpose it is desirable to make the *CSVs* as point-specific as possible and provide sufficient classes to discriminate among values in sets of neighboring vertices, in order to pick points that best meet the selection criterion used. This criterion — which we call the *preferred sinuosity value (PSV)*, parameter 8 in the above list (called *hint* in pseudocode listings) — is an integer between one and the number of sinuosity classes encoded. After *weedQids* has sampled a run of vertices to retain one or more of them, the *CSV* of each selected one is compared to that of its immediate neighbors to determine if any of them are closer to *PSV* than the already-selected vertex. Should one have a closer value, that vertex will be selected for output instead of the one chosen mechanically. If several such candidate vertices are found (which will happen more often when there are few sinuosity classes), the one topologically closest to the already-selected one is selected. This tie-breaking rule tends to maintain a fairly even spacing between points, since they were already systematically sampled.

In re-selecting points, the caller can cause low, high or intermediate values of *CSV* to be preferred, according to the specification of *PSV*. Points of high sinuosity are normally preferred, as they are visually more significant than others. However, in complex linear features, it can help to select points of lower sinuosity, to quiet the region a bit and provide a diminished opportunity for self-crossing. As many generalization problems are context-dependent, using a single *PSV* may be sub-optimal, even if it is independently re-specified for each primitive. This need to fine-tune sinuosity handling can be filled by computing sinuosity in such a way as to characterize sections of arcs rather than individual vertices.

Line Characterization. Various researchers have investigated ways to improve map generalization by segmenting features into similar portions that are then generalized separately, using different parameter settings (Buttenfield 1986, 1989; Muller 1986; Plazanet 1995). Local measures of curvature and dimensionality are frequently used to do this, and methods can be local (e.g., linked bi-cubic splines) or global (e.g., Fourier transforms). We use *SVs* for this task, computing them so that minor variations from point to point are smoothed over, maximizing the number of contiguous vertices with the same *CSV*. In this approach, we normally observe the following rules: (1) larger neighborhoods (greater values for m and n) are used, up to about seven vertices; (2) SVs are mapped into fewer classes (three values for *CSV* seem enough); (3) the *CSVs* are then smoothed (sent through a low-pass filter) to alias short sinuosity regimes to the *CSV* of one of their neighboring runs. Figure 4.10 describes and illustrates this filtering process.

Real measures of sinuosity for line vertices may be classified into a small number of regimes, for example to identify regions of low, medium and high line sinuosity. This line has been classified into 3 CSV levels, yielding the following results:

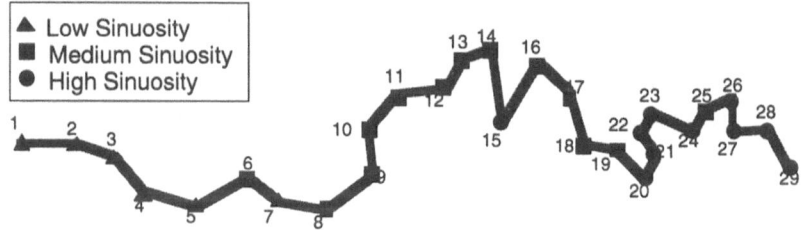

▲ Low Sinuosity
■ Medium Sinuosity
● High Sinuosity

Points 6, 15 and 25 are dissimilar to their neighbors and fall into a different class. While this reflects the line's structure, it may not be useful in analysis to have the sinuosity of one or two points differ from those of their neighboring ones.

PNT. NUM.	GIVEN CLASS	SM. SIN. CLASS
1	1	1
2	1	1
3	1	1
4	1	1
5	1	1
6	2	1
7	1	1
8	2	2
9	2	2
10	2	2
11	2	2
12	2	2
13	2	2
14	2	2
15	3	2
16	2	2
17	2	2
18	2	2
19	2	2
20	3	3
21	3	3
22	3	3
23	3	3
24	3	3
25	2	3
26	3	3
27	3	3
28	3	3
29	3	3

SMOOTHED CLASSIFIED REGIMES

This filtering can be performed across any number of vertices; weighting schemes may be used if desired. We employ a single parameter, the smallest number of consecutive vertices of any given class that will be tolerated as a run. Runs of vertices having fewer members will have all their CSVs aliased to one (or both)of their neighboring runs, giving precedence to higher values of sinuosity if necessary.

In the above example, it is sufficient to set the smoothing parameter to two, meaning that only individual deviations from neighboring runs will be aliased. Usually the parameter needs to be set higher, typically to four or above.

The table to the left shows the effects of filtering out erratic single values, with the modified classification shown above.

Fig. 4.10: Filtering Classified sinuosity Values to identify homogenous regimes

We characterize lines in order to tailor generalization parameters to them according to their geometric complexity. This may also be done according to non-geometric attributes such as feature class, but no variables other than classified sinuosity were used in this study. It was decided to limit such segmentation to three sinuosity classes, as they should be sufficient to test the benefits of this approach, and any more might be of spurious utility. In these tests, runs of fewer than three vertices with the same *CSV* were merged by the smoothing operator with adjoining runs, promoting their *CSV*s in preference to demoting them. The classified primitives were then output in homogenous segments, one class at a time, for further analysis. Findings from these studies and strategies built on them are presented in chapter 5.

4.5   Chapter 4 Summary

Several ways in which QTM-encoded map data can be generalized were discussed and categorized as (1) simple detail filtering for individual map objects, (2) detail filtering for single features using attractor occupancy, and (3) spatial conflict negotiations for multiple features using attractors. The first category was seen to have limited power, but provides a strong platform for the other two. A flexible algorithm for the second category was described and aspects of it illustrated; some of these make heuristic use of "enriched" QTM IDs to help decide which vertices to select when changing scale. After elucidating a set of eight QTM generalization parameters, certain computational considerations were discussed, and pseudocode to implement the parameters was presented.

The generalization approaches explored in this chapter are supported by a feature-based data schema which can use any information concerning topological relationships in a database, but does not require it. The model allows features to be designated as collections of primitive elements (points, arcs, polygons). An element may play a role in more than one feature (be a "shared primitive") and may be topologically linked to other primitives. Topology and sharing can be exploited to perform generalization, as they both indicate what other elements may be affected by geometric alteration of a given one. In this data schema, nearly everything is a list, either of features, primitives, vertices or attractors, although other data structures are also used. For identifying spatial conflicts among features, directory structures were described that relate primitives and attractors they occupy.

Viewing data through the above schema, problems of generalizing multiple features and feature classes that compete for space were discussed next. While QTM does not offer any magical solutions, it is capable of providing a framework within which multi-feature generalization strategies can be developed. This framework is the space-primary mesh of attractors, which may be used to identify competing features at any chosen level of detail. General guidelines for how to select an appropriate detail level were offered, and this problem was seen as related to the method used to index attractor data. Several indexing schemes, including trees, linked lists, hash tables and directories, were discussed. Which of these is optimal is probably application-dependent and may be constrained by memory resources. Nevertheless, the directory-based methods described subsequently seem useful in fulfilling present objectives.

In the preceding discussions, it was assumed that attractor data would always be ephemeral (generated on-the-fly), and would not be committed to a primary spatial database. However, *results of analyzing attractor occupancy could be stored in a database* in interpreted form. This auxiliary data could describe pairs or groups of features that required negotiation at specific scales (QTM levels of detail), without necessarily specifying the outcome of those negotiations (which could change depending on the contents and purpose of the generalized maps). This would not add much bulk to one's database, but could drastically reduce the search space needed to generalize across feature classes on later occasions.

The chapter concluded with a description of a technique for adding geometric metadata (*sinuosity hints*) to QTM-encoded features within QTM identifiers describing vertex locations (a *data enrichment* technique). Two ways in which this information can be used for generalization were discussed, one intended for steering vertex selection during simplification, the other for segmenting primitives into subsections having more uniform sinuosity characteristics. Initial results of using sinuosity measures for both purposes, along with other effects of varying QTM generalization parameters, are presented in chapter 5.

5 QTM Scale Filtering: Tests and Evaluations

One person's data is another person's trivia
— Anon.

Having explained *how* QTM map generalization can work, we are ready to look at *how well* it can work. By this we mean not only whether currently-existing QTM-based generalization operators function properly and reliably, but also how the results of their application to real-world data measure up, how aesthetic they are, and what could be done to improve them.

We start with a quick overview of space-primary generalization strategies as implemented by several researchers. Before moving on to our empirical results, the environment and test data used are described, including the components of our testbed and some details concerning modeling and handling test data. Two simple datasets were used in these cases, both shorelines of islands. The rationale for using these examples is articulated, and the test methodology explained (detailed tabular and graphic results from the tests are provided in appendices C and D). This is followed by an evaluation of results, in which the qualitative and quantitative consequences of applying these algorithms are discussed in an attempt to dissect effects of input parameters.

In the course of presenting the details of empirical findings, assessments of the apparent strengths and weaknesses of our methods are offered. We make some comparisons with a well-known line simplification procedure, which leads to some ideas concerning why certain results occurred, how more complex situations might be handled, and what might be needed to make our and other generalization methods function better. Additional discussion about generalization strategies and plans for future research may be found in chapter 6.

5.1 Evaluating QTM Line Generalization

The methods of line simplification explored in this study are novel, but are typical of *space-primary* approaches. The portion of this repertoire that performs line simplification almost exclusively deals with raster map data. One of several exceptions is an algorithm by Li and Openshaw (1993) which was implemented in both vector and raster modes (described in Li and Openshaw 1992). The approach is based on the "natural principle" of resolving power, mimicking how the eye gracefully degrades detail as objects recede in the distance. It uses "epsilon band" spatial filtering (Perkal 1966), where the widths of bands are determined not by positional uncertainty of the data but by the target spatial resolution.

More recently, Zhan and Buttenfield (1996) describe a technique for encoding vector source data for display in the form of a resolution pyramid having up to 10 levels of quadtree pixel fields. Cartographic line data are scaled to the size at which they will be displayed, then sampled at a fine resolution (0.2mm is used). The samples are chained in sequence and quadtree leaf codes are computed for them. Resolution is then recursively cut in half, and locations in "parent" (lower-resolution) quadrants are computed to represent pixels in their child quadrants, using algorithms that merge and smooth pixel details. Graphic results are quite good, even though a number of lower-resolution stages look about the same for some examples. The authors claim that lines thus become *multi-resolution cartographic objects,* but they fail to point out that such "objects" are rooted at a particular display scale (whatever ground distance 0.2mm is presumed to represent) and would need to be rebuilt if these conditions change. One also wonders how efficient the encoding, storage and decoding of these objects might be, which the paper did not addresss. Nevertheless, there are some strong similarities between this approach, that of Li and Openshaw and our own.

Space-primary generalization techniques such as the above two examples partition space rather than phenomena to decide how and where to simplify details. They identify what portions of features cluster in pre-defined chunks of two-dimensional space, and then modify attributes of the carved-up space until only one simple description (usually of a single feature) remains in any one place.

The algorithms developed so far do little more than this, but they hint at greater possibilities. Much of their potential derives from the fact that although space-primary techniques are used, it is *object-primary data* to which they are applied (see fig. 2.1). This is a relatively unexplored area, a potential synthesis of data models long-sought by both the GIS and image processing communities (Peuquet 1984), that may also answer some of the objections raised in the GIS community about the appropriateness of hierarchical approaches (Waugh 1986).

Given the novelty of our approach, it was deemed most important to understand as well as possible the effects of controls to QTM-based line simplification before attempting to apply it to complex cartographic situations. The tests reported below are an effort to do that; hopefully they provide a sound basis for refining our techniques. Certain results of the tests are presented below, following explanations of the software and hardware environments and descriptions of the test datasets. Empirical tabular results are presented in appendix C, and a subset of these are shown graphically in appendix D.

5.1.1   Software Testbed

Implementation of the generalization methods described in chapter 4 entailed development of software to read vector map data from files, converting coordinates to QTM identifiers and structuring data internally using the data model described in section 4.4.1. This testbed was named *QThiMer* (for *QTM hierarchy*), and was written in C using the Metrowerks *CodeWarrior* integrated development environment, v. 9. *QThiMer* consists of about 6,000 lines of commented C code defining around 200 functions, and requires about 1 MB of main memory to execute, slightly more for large datasets (those exceeding about 250K of ASCII source coordinate data). Data structures, which are generally object-oriented, could be re-engineered to lower memory requirements. In general, optimizing our code was not a major priority.

All work was carried out on an Apple Macintosh™ *fx* system (33 MHz 68040 CPU, 32MB RAM) running *MacOS* 7.5.3. As this machine is an obsolete platform, it was decided not to benchmark processing times for runs. Creation of a *PowerPC* version of the testbed is a goal for the future, at which point timing data will be collected and analyzed. On the Macintosh *fx*, rough measurements indicated that the core conversion function (*qtmEval*) consumes roughly five milliseconds per coordinate, but this depends on the QTM depth of conversion and other factors (see appendix A).

Once installed in a memory-resident hierarchy of primitive, feature and feature set data structures as shown in figures 4.5 and 4.6, these objects were subjected to QTM detail filtering using pre-defined sets of generalization parameters, and results were output as ASCII files of geographic coordinates via inverse transformations from QTM IDs to latitudes and longitudes.

Input and output data were analyzed by *QThiMer* to compute basic statistics, such as polygon area, arc length, average segment length and minimum bounding boxes and triangles, for use in subsequent analyses. To display results, *Atlas Pro GIS*, v. 1.08 was used, mainly for its support for input of ASCII coordinate data. Due to that application's quirks and limitations, map graphics from *Atlas* needed to be transferred to a drawing program to be made presentable. *Canvas* v. 3.5 handled this; data was communicated to it in the form of PICT files, which were resymbolized, rescaled, annotated and printed directly by *Canvas* on a 600 DPI laser printer for inclusion in appendix D.

We would have preferred *QThiMer* itself to handle these graphic manipulations, which would have eliminated numerous annoying tribulations of handling intermediate files. Building an interactive GUI (graphic user interface) and coding graphic support functions would also have speeded data analysis and enabled more elaborate experiments to be conducted. Given that the development time frame was quite short (only 18 months in total), and given our desire to build a platform-independent prototype, it was decided not to spend time constructing a GUI, even though this meant *QThiMer*'s ease-of-use would suffer. Consequently, all interaction with *QThiMer* is via prompts issued and commands typed to a terminal window. Reports from the program describing data objects and processing results are sent to that window or redirected to text files for further analysis. While the prototype software is inelegant from a useability point of view, we feel that the time that GUI design and coding would have required was far more profitably spent in refining algorithms and data structures and studying their behavior.

5.1.2 Test Data

One of the goals for conducting empirical tests was to simply see how reasonable the results of QTM-based line simplification might be from a purely cartographic point of view. To do this, we felt it would be best to find map data which encode the same set of features at different scales, such as might appear on different map series published by a national mapping agency (NMA). The actual location and origin of such data was seen as less important than having multiple representations be available, preferably of curvilinear natural phenomena rather than of more rectilinear man-made features. We also wished to test several different feature classes, in order to assess whether some are more amenable to our methods than others. However, time pressure caused us to put this goal aside, and consequently only one class of data was studied.

Beyond the Swiss political boundaries displayed in chapter 3, there were little map data available in-house suitable for these tests, which led to a search being made on the internet. Our attention was directed to an WWW server operated by the US Geological Survey, called *coastline Extractor*.[1] The forms on that page allow one to specify one of four source scales, a retrieval window (in latitude and longitude) and an output format for extracted coordinates (we used ARC/INFO *ungenerate*). The datasets provided contain only coastline shapes, without any geocodes, metadata or attributes, drawn from the following U.S. sources:

1. World Coastline Database (U.S. NOAA, ca. 1:5,000,000)

2. World Data Bank II (U.S. CIA, ca. 1:2,000,000)

3. World Vector Shoreline file (US NOAA, ca. 1:250,000)

4. Digital Nautical Charts (U.S. NOAA, 1:80,000 - 1:20,000)

The first two datasets are quite small-scale, and are not very suitable for our purposes, but were useful for reference. The third class (WVS) of data also provides world-wide coverage, at a scale that is marginally useful. These datasets were created by scanning remotely-sensed radar images and vectorizing the land-water boundaries; as a result it has a jagged appearance when viewed at scales of 1:250,000 or more. This can be improved by line smoothing operators, and this was sometimes done. Almost all the tests were performed on category 4 data, the digital nautical charts. As the above list indicates, these were digitized from published charts at several scales, the base scale being 1:80,000. However, some larger-scale map insets were also digitized and included in the data stream without identification of these scale transitions. Furthermore, these data include only U.S. coastal waters, which restricted the locus of the tests. But as there was no requirement to study any particular part of the world, this was not an impediment.

We felt it was important was to find data for map features that had a range of characteristics that would pose specific problems to generalization, such as the

---

[1] http://crusty.er.usgs.gov/coast/getcoast.html

potential for self-intersection. It was also necessary for features not to be too large, in order to be able to display them at close to source scale on letter-size paper. While extracts from larger features could have been used, we wished to avoid this in order to preserve the gestalt of the features while still displaying their details. For this reason, we selected *island* outlines to work with, rather than clipping portions from mainland coasts.

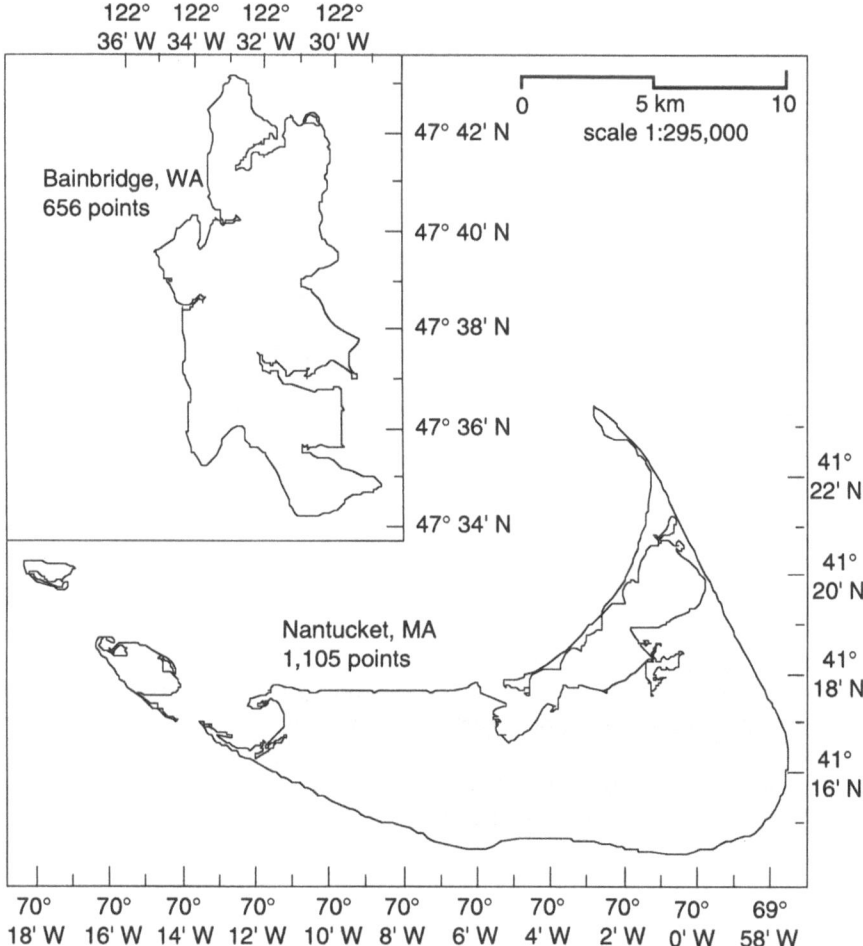

Fig. 5.1: Bainbridge and Nantucket Islands 1:80K source data (ungeneralized) displayed at 1:295K

One of the two test areas chosen is an island off the east coast of the U.S., the other from its west coast. The former is Nantucket County, Massachusetts, a group of one large and two small islands about 40 km south-east of Cape Cod in the Atlantic Ocean. These islands have no bedrock, being stabilized sand, sculpted by waves and tides. The other example is Bainbridge Island, located five thousand km away in Puget Sound, about 10 km west of Seattle, Washington. It is in an area of more complex geology, with an undulating, rocky shoreline punctuated by deep tidal inlets. Basic statistics for the two datasets are presented in table 5.1.

Attribute	Nantucket MA	Bainbridge WA
Area (km^2)	125.02	70.00
Perimeter (km)	151.92	77.42
N/S Length (km)	17.06	16.41
Digitizing Scale	1:80,000 or more	1:80,000 or more
E/W Length (km)	29.54	8.59
Avg. Seg. (km)	0.14	0.12
No. of Vertices	1105	656

Table 5.1: Statistics for Nantucket MA and Bainbridge WA Test Datasets

Figure 5.1 brings the two datasets together at the same scale. Except for a small cape on its north-east extreme, Bainbridge does not have the prominent spits and bars that characterize Nantucket, and Nantucket does not have the kid of deep-cut estuaries found along the Bainbridge shore.

5.1.3   Methodology

Datasets were obtained from the USGS *coastline Extractor* web site, described in section 5.1.1, in ARC/INFO *ungenerate* format. The shorelines had been captured as arbitrary arcs, and the ones comprising the islands of interest were identified within *Atlas Pro*; new source files were produced that contained just these arcs. Each of the two files was pre-processed separately by *QThiMer* to link connected arcs into polygons; the resulting files were then repeatedly processed by *QThiMer* to generalize the coastlines using different algorithms and parameter settings.

Two line simplification algorithms were employed: (1) the *weedQids* algorithm, described in appendix A, and (2) a stack-based version of the Ramer-Douglas-Peucker (RDP) algorithm for comparison. When such comparisons are made, the QTM simplification is done first, then the latter algorithm applied to yield the same number of retained vertices, selecting a band tolerance which yields this result. These subsets of the original points are output as coordinate text files in Atlas BNA format. These were converted by *Atlas Pro* into Atlas "geographic" files, transforming coordinates into one of three supported map projections; we elected to use a simple rectangular (platte carré) projection. In *Atlas Pro* these projected files were organized into several related thematic layers representing a sequence of simplifications; some times parameters would vary, other times levels of detail or algorithms, depending on our purpose and type of comparisons we were attempting to make. Once the files were installed as *Atlas Pro* layers, they could be colored, shown or hidden prior to being exported as Macintosh *PICT* files. The *PICT* files were read by *Canvas*, which was used to scale, symbolize, annotate and format graphics for display in this document.

In addition to repeatedly generalizing the coastline data, *QThiMer* was used to analyze coastline sinuosity, using the algorithmic approach described in section 4.4.3. Local sinuosity at each vertex in a dataset was estimated and assigned to one of three classes, called *low*, *medium* and *high* using the $\sqrt{(1-1/SV^2)}$ classifier described in fig. 4.8, followed by a smoothing of sinuosity values that eliminated runs of less than four vertices. *QThiMer* was modified to enable vertices to be output as separate files, one for each class. This enabled further analysis and graphic display of the classified coastlines. As only three classes were used, and because some vertices were coerced into another class by the smoothing operator, the results of this process are highly approximate. They are still, however, suitable for investigations of the response of cartographic lines to generalization operators.

Basic statistics describing generalization results were computed by *QThiMer* and transferred to Microsoft *Excel* v. 5, where summary tabulations, derived measures, statistical tests and charts were made. Most of the tabular data are provided in appendix C, along with selected plots showing relationships between measures.

5.2   Empirical Map Generalization Experiments

As just described, several shoreline datasets were processed in parallel in exploring the effectiveness, properties and limits of QTM-based map generalization. Each of the two study areas were treated identically, generating output data that are summarized in appendices C and D. Therefore, the contents of figures D1.1 - D 1.12 for Nantucket Island are mirrored by figures D2.1 - 2.12 for Bainbridge Island, one for one. Readers may wish to visually compare respective figures from the two sets. Also, within each set, there are sequences of figures that are meant to be looked at together; these are D1.2 - D1.7, D1.8 - D1.9, and D1.10 - D1.12 — and similarly D2.2 - D2.7, D2.8 - D2.9, and D2.10 - D2.12.

Source Data. Figures D1.1 and D2.1 show the (ungeneralized) source data used in experiments, drawn at the scales indicated by their provenance. Also displayed is the QTM grid at level 12 in the hierarchy, which is characterized by facets with 1 km edges and 1 km^2 area. This is one level below the coarsest resolution reached in our generalization experiments, and is intended to serve as a guide for interpreting graphic results. Notice that the facets are shaped differently in Washington State than they are in Massachusetts. While the two states have fairly similar latitudes, they are about 50 degrees apart in longitude, more than enough to make noticeable differences in the tessellation's geometry. Although one might wonder about possible consequences, there is no inherent reason to believe that these changes in shape affect the reported results.

Generalization Controls. Figures D1.2 - D1.7 and D2.2 - D2.7 display results of QTM generalization under controlled conditions. Each shows a progression of simplification through six halvings of scale (1:95K to 1:1.3M) — QTM levels 18 through 13, respectively. Each figure documents the parameter settings used, as well as some basic statistics for each of the generalizations.

Controlled Parameters. The parameters used in each experiment (discussed in section 4.4.2) are reported in D2-D7 as:

Mode:	Whether the generalization process was conducted hierarchically or non-hierarchically.
Mesh:	The type of mesh element used to filter detail; either QTM facets or QTM attractors.
Lookahead:	How far runs of QTM IDs are examined to identify a return from an originating mesh element.
Points/Run:	Maximum number of vertices that will be retained to represent a run within each mesh element. In these tests, only one point was chosen per run .
Accuracy:	The linear uncertainty of positions of vertices when decoded from QTM IDs to geographic coordinates. All our tests position points using the maximum precision available for them.
Sinuosity:	The degree of local sinuosity preferred when selecting points for output; expressed as class *n/m* classes. However, in figures D2 through D7, points were selected mechanically, without regard to their sinuosity indices.
Line Weight:	The point size of lines used to represent coastlines, usually 0.5 (about 0.2 mm); fine hairlines are used in some insets.

Generalization Statistics. A table included in figures D2 through D7 summarizes the data shown in each of the six maps shown. The columns in these tables describe:

LVL:	The QTM level of detail used for filtering; refer to table 2.1 for more general properties of QTM levels.
PNTS:	The number of polygon vertices retained by the simplifications.
PCT:	The percent of original points retained by the simplifications.

LINKM: Linear measure (length) of generalized coastlines in kilometers.

SEGKM: Average segment length (distance between vertices) in kilometers, computed as LINKM/(PNTS-1).

SEGMM: The average length of segments (in millimeters) on a map display at the "standard scale" for each QTM level of detail, computed by multiplying SEGKM by 1,000,000 and dividing by the scale denominator.

The first three statistics are self-explanatory. *LINKM* is a rough indicator of shape change; it always declines with the level of detail, and when it changes greatly so usually do some aspects of the features' configuration.

Generalization usually reduces the number of map vertices much more than its perimeter shrinks, and this is reflected by the *SEGKM* statistic, which tends to monotonically increase as levels of detail decrease, and can change by an order of magnitude across the scales. Because it is derived from a filtering process using QTM mesh elements. it is correlated with linear QTM resolution, but does not vary linearly with it.

The *SEGMM* statistic is an interesting and useful indicator of one aspect of generalization quality. Unlike *SEGKM*, which tends to increase with simplifica-tion (inversely with QTM level), *SEGMM* varies much less. It often decreases slightly at smaller scales, but can remain constant or increase as well. *A "good" generalization method should yield values for SEGMM that do not vary with scale change, as this would indicate that the amount of mapped detail remains constant at all scales.* An optimal value for *SEGMM* is hard to pin down, as it somewhat depends on source data density, graphic display resolution and map purpose. However, for the sake of consistency, *we have made the assumption that an optimal value for SEGMM is 0.4 mm*, the same minimal distance we use for de-riving map scale factors from QTM levels of detail. This is about twice the line weight used in most illustrations in appendix D.

Using *SEGMM* as a quality metric for the maps in figures D2-D7 tends to con-firm this, although fine distinctions are not easy to make. In general, when *SEGMM* falls below 0.4 mm, the amount of line detail tends to clog the map and overwhelm the eye, and serves no useful purpose. When *SEGMM* rises above 1 mm (which it does only in a few instances), the graphic quality becomes notice-ably coarser, and using less filtering may be indicated. The most pleasing results seem to be found when *SEGMM* is in the range of 0.4 to 1 mm. This is most of-ten achieved by using attractor-driven (rather than facet-driven) generalization, and a hierarchical (rather than non-hierarchical) strategy, but even some of these results suffer in quality.

Comparison of Generalization Results.   Figures *D8 and D9* provide di-rect comparisons between QTM and RDP simplification and other source data. For convenience, a portion of fig. D1.8 is reproduced as fig. 5.2 and D2.8 is also excerpted in fig. 5.4. Three levels of simplification are displayed, comparing results from (1) QTM facet, (2) QTM attractor, and (3) RDP bandwidth-tolerance methods. Note the displays of filtering elements to the upper right of each of the nine plots, drawn at scale (the smaller RDP bands are almost invisible, however).

The *D9 figures* provide a different type of comparative evaluation, in which source data at several scales may be compared to QTM-generalized versions of them. This figure does not really compare automated to manual generalizations, as the source data come from independent digitizations of different map series at different scales. The top row of maps show the data we have been using, the 1:80,000 NOAA/NOS digital navigation charts. The lower row shows World Vector Shoreline Data and World Data Bank II shorelines (the latter is displayed for comparison only and was not generalized). Moving from left to right, source files are generalized (using a non-hierarchical attractor methodology) to 1:250K and 1:2,000K, as the column headings indicate. Comparisons should be made between same-scale maps found along minor diagonals (bottom-left to upper-right), and between the two maps in the rightmost column. Footprints of source data are shown in grey to assist in evaluating these results.

Like figures D9, the *D10 figures* show the 1:250K WVS data, and compare its simplifications to WDB II. A two-way comparison is made between the effects of specifying *high sinuosity/no sinuosity* and *two-point lookahead/no lookahead*. Level-14 generalizations are appropriate for 1:1,000K display; the single level-13 map (at lower left) is appropriate for 1:2,000K display, as the white insets in the grey source data footprints illustrate.

Figures *D10 - D12*, illustrate some properties that measures of line sinuosity can contribute to the generalization process. D11, a multiple of 17 maps plus 8 insets, shows the effects of specifying a "preferred sinuosity value" (PSV) when selecting vertices to be retained by QTM line simplification (see discussion in section 4.4.3). Three QTM levels are shown. Using both hierarchical and non-hierarchical strategies, preference is given to (1) low-sinuosity vertices or (2) high-sinuosity vertices, (1 of 7 and 7 of 7 classes, respectively). The resulting distribution of sinuosities are quite different, as histograms in fig. *C2* portrays. The maps reveal a number of marked differences caused by both the strategy and the sinuosity selected, but these tend to manifest themselves only at higher degrees of simplification. Choosing vertices of low sinuosity does tend to eliminate some spikes and superfluous detail for highly-simplified features, just as using a hierarchical elimination strategy does. That is, the *hierarchical/low sinuosity* combinations tend to work better at small scales.

Figures D12 depict the results of classifying measures of sinuosity for map vertices rather than offering comparisons. Although seven levels of sinuosity were employed in previous examples, only three classes are used here, and these were subjected to an attribute smoothing procedure. We did this because, as discussed in section 4.4.3, a smaller number of class distinctions aides in partitioning datasets, whereas a larger number is more useful in selecting among vertices. Here the goal is to segment features in such a way as to identify homogeneous stretches of coast. Having done this, different generalization parameters (or even different operators) could be used in each regime separately, to tune the process even more closely. The D12 maps show three classes of sinuosity, symbolized with light grey, dark grey and black lines. They also contain 9-class histograms of vertex sinuosity values, shaded to group the bars into the three classes mapped. Although there are a fair number of high-sinuosity points, their line symbolism is not very extensive, due to the closer spacing of such vertices.

5.2.1   Results for Bainbridge Island, WA

Section 4.2.2 explains the difference between attractor- and facet-based QTM line simplification. Figure 5.2 compares results of the *weedQid* and *RDP* algorithms for Bainbridge (see D1.8 for a full illustration). It shows QTM filtering by facets and by attractors at three levels of detail (14, 15 and 16), in comparison to RDP filtering, performed so as to yield the same number of points (or very close to that) which resulted from using attractor filtering.

Note the triangles and hexagons to the right of each map in the first two columns. These show the extents of the QTM facets and attractors through which filtering was performed. The third column displays the bandwidths used for RDP filtering, but barely discernible in the top row. As one might expect, using attractors as mesh elements always causes more simplification, here changing by less than a factor of one-half; a less rugged feature would have received more simplification using attractors, other factors being equal. Also, as often seems to be the case, attractor filtering results in about as much reduction as when facet filtering is performed at the next lower level; compare the top two figures in column 2 with the bottom two in column 1. Despite equivalencies in the number of points, noticeable differences appear to result from applying different filtering elements, with attractors tending to produce a more coarse result (which often may be more useful). Bear in mind that attractors are both hexagonal and triangular, the former being six times larger than the latter. This prevents the coarseness of filtering from becoming as great as the attractor symbols on fig. 5.2 would tend to indicate.

Comparing attractor filtering with Ramer-Douglas-Peucker results yields a similar impression: RDP tends to look less refined than QTM attractor filtering when the same number of vertices are retained. Using RDP, coastlines tend to get straighter (but not necessarily smoother), often looking more angular. Overall, RDP seems to exhibit less consistency, possibly because it need not sample vertices the same everywhere, as QTM-based methods do. Comparable experiments by Zhan and Buttenfield (1996) produced similar results in comparing RDP to their space-primary method.[2] Furthermore, results from RDP tend to be sensitive to the choice of the initial endpoints, and in this and other cases studied here they are the same point, the features being defined as polygons. In this study, we did not explore changing the endpoints, although that would naturally result were we to have generalized sinuosity-classified features such as are shown in figures D12.

Does Simplification Alter Line Character?   A different use of sinuosity statistics than figures D11 (selecting vertices) and D12 (segmenting features) show was also explored. We wished to determine if any of the generalization methods we used altered the overall distribution of vertex sinuosity of features being simplified. To assess this, we computed three classes of vertex sinuosities, as in D12, but *did not* smooth the results, as the aim was to classify the importance of vertices rather than to segment features. Both input and output datasets were tabulated, comparing the distributions of sinuosity values across five levels of detail for the three simplification methods illustrated in fig. 5.2, plus attractor-filtered vertices in which high values of sinuosity were favored during selection.

Results of this analysis are displayed as histograms in fig. 5.3. A chi-squared test was applied to assess if any of the four methods altered the distribution of sinuosities found in the input data as a side-effect of line simplification. As the black dots in fig. 5.3 indicate, similarity with the original distribution of sinuosity was maintained only in the one, two or three most detailed levels, and the RDP algorithm did the best job at doing this. At lower levels of detail, QTM-based methods tended to select higher-sinuosity vertices than did RDP. This may partially explain some of the differences in line character observed in fig. 5.2.

How important might it be to maintain a given distribution of vertex sinuosities when simplifying linear features? Is there any relationship at all between our quantification called *sinuosity* and the graphic quality of generalized features? Figure D1.11 (and its companion chart C1.2) try to sort out the effects of specifying low and high sinuosity values as heuristics for selecting vertices to be retained. In these examples selecting low-sinuosity points appear to produce better results, in the sense that some spikes are avoided and narrow passages tend to be opened up. This makes some intuitive sense, but note that it is almost opposite to the behavior of the RDP algorithm, which tends to select locally extreme points, sometimes creating jaggedness. We also note that, sinuosity-based selection has a stronger effect at higher degrees of simplification; the more simplified maps in D1.11 (those at the top) are clearly improved by using low-sinuosity selection, while the more detailed maps at the bottom are rather less affected (but the low-sinuosity versions seem to be slightly superior as well). We address this issue in greater detail in section 5.3.

---

[2] The Zhan-Buttenfield algorithms perform an explicit smoothing operation, having the effect of displacing retained points from their original locations, while both RDP and weedQids simply select points to be retained. This could account for observed differences in simplified line character.

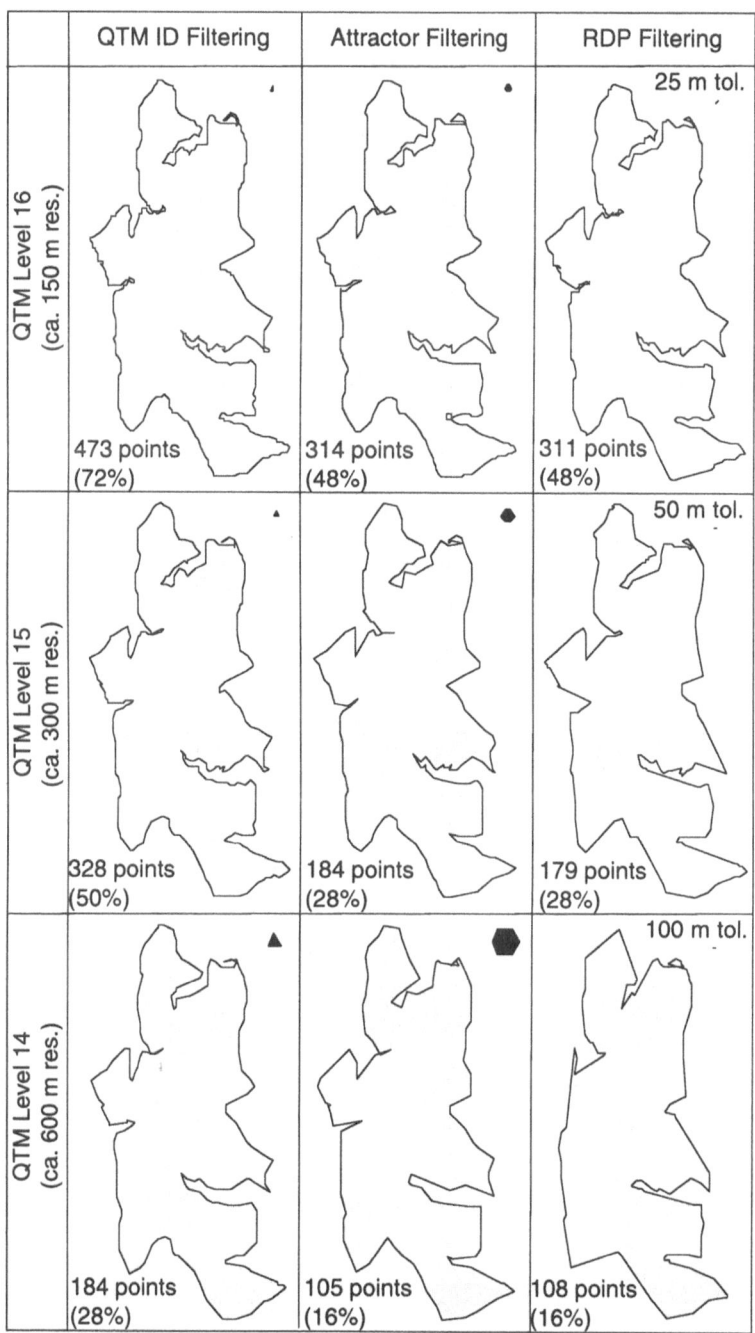

Fig. 5.2: Bainbridge Island WA: Three simplification methods compared

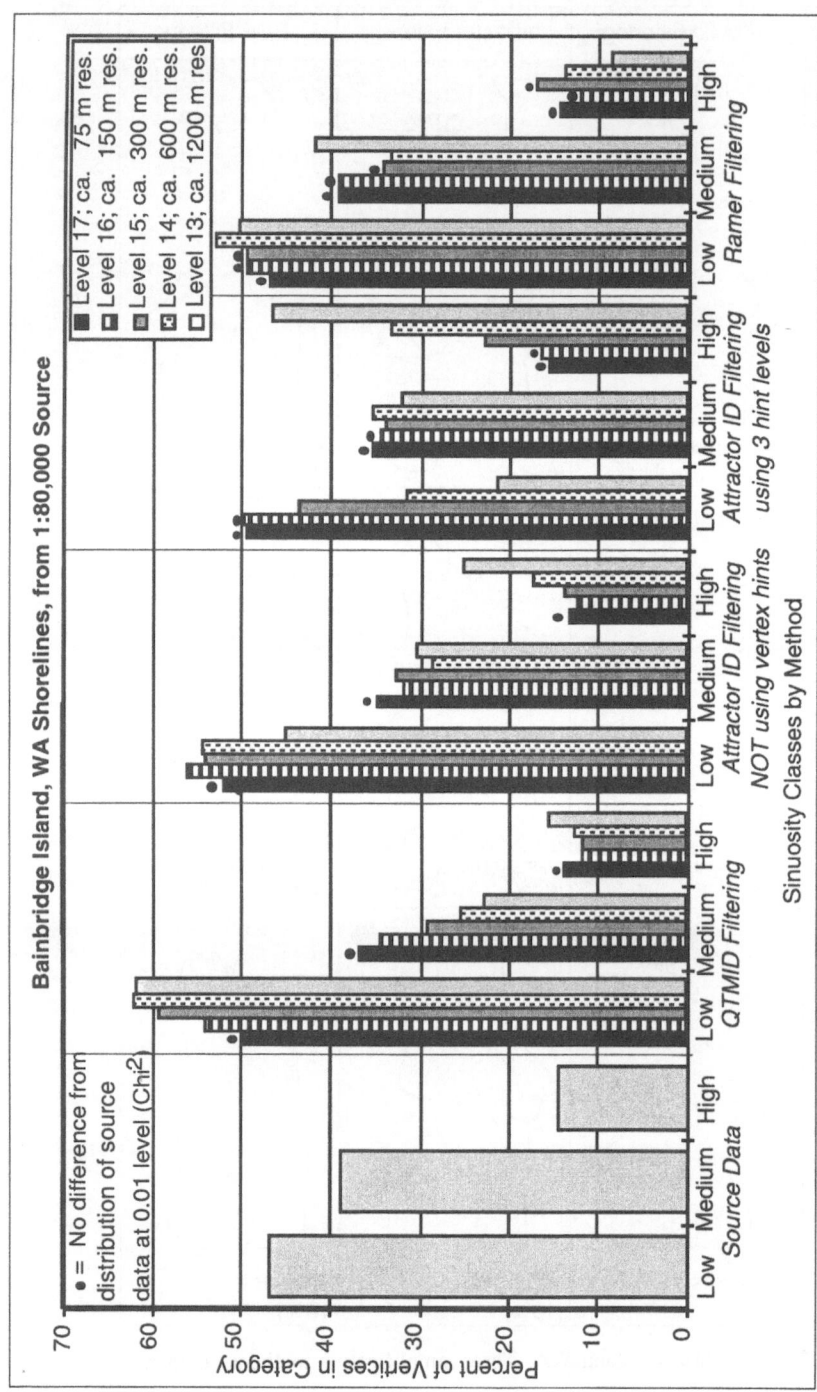

Figure 5.3 Relative distribution of three categories of Vertex Sinuosity before and after applying four generalization methods

5.2.2   Results for Nantucket Islands, MA

Nantucket lies off the Southeast coast of the state of Massachusetts and is somewhat larger than Bainbridge. It also has a completely different character, essentially being a shoal — a sand-pile laid upon the continental shelf. While they have some sharp features, including sandbars, spits, harbors and estuaries, these islands are essentially a crescent of smooth, sandy shores formed by wave action; the height of land on the main island is less than 20 meters above sea level. Their shorelines (as delineated from 1:80,000 data and displayed at 1:125,000) are best represented in fig. D2.1, with facets of the 12-level QTM graticule superimposed.

Because they are already rather spike-like, the islands' sand-spits are difficult to generalize by any automated method. Simplifying them tends to thin them into nothingness, or coagulate into lumps lacking cohesion. Properly generalizing such features really demands the use of *displacement operators* to progressively widen the sand-spits as scale diminishes. At some point, bars would be enlarged to occupy more the bays behind them.

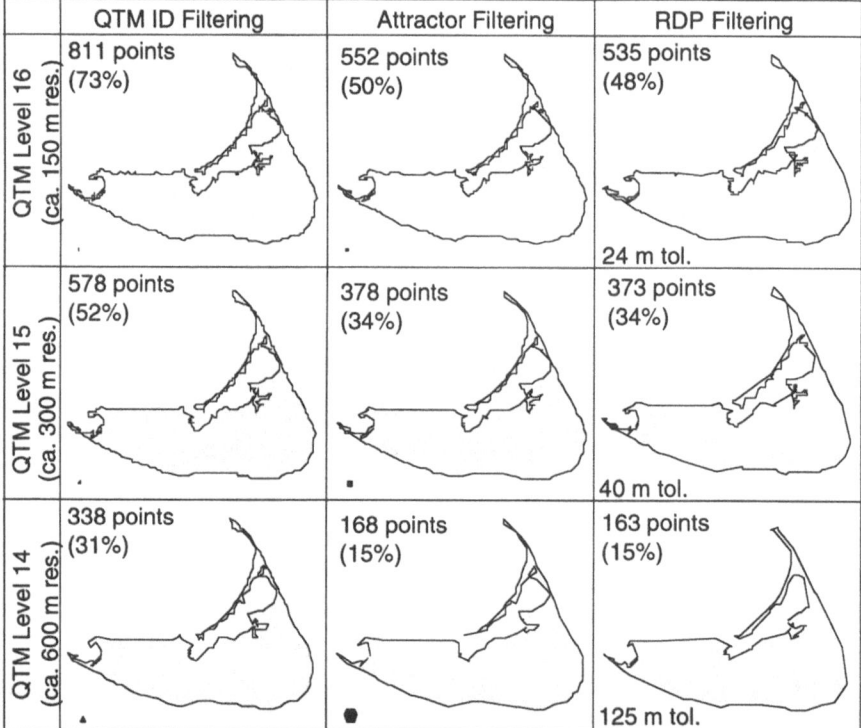

Fig. 5.4:   Nantucket MA: Three simplification methods compared (minor islands excluded)

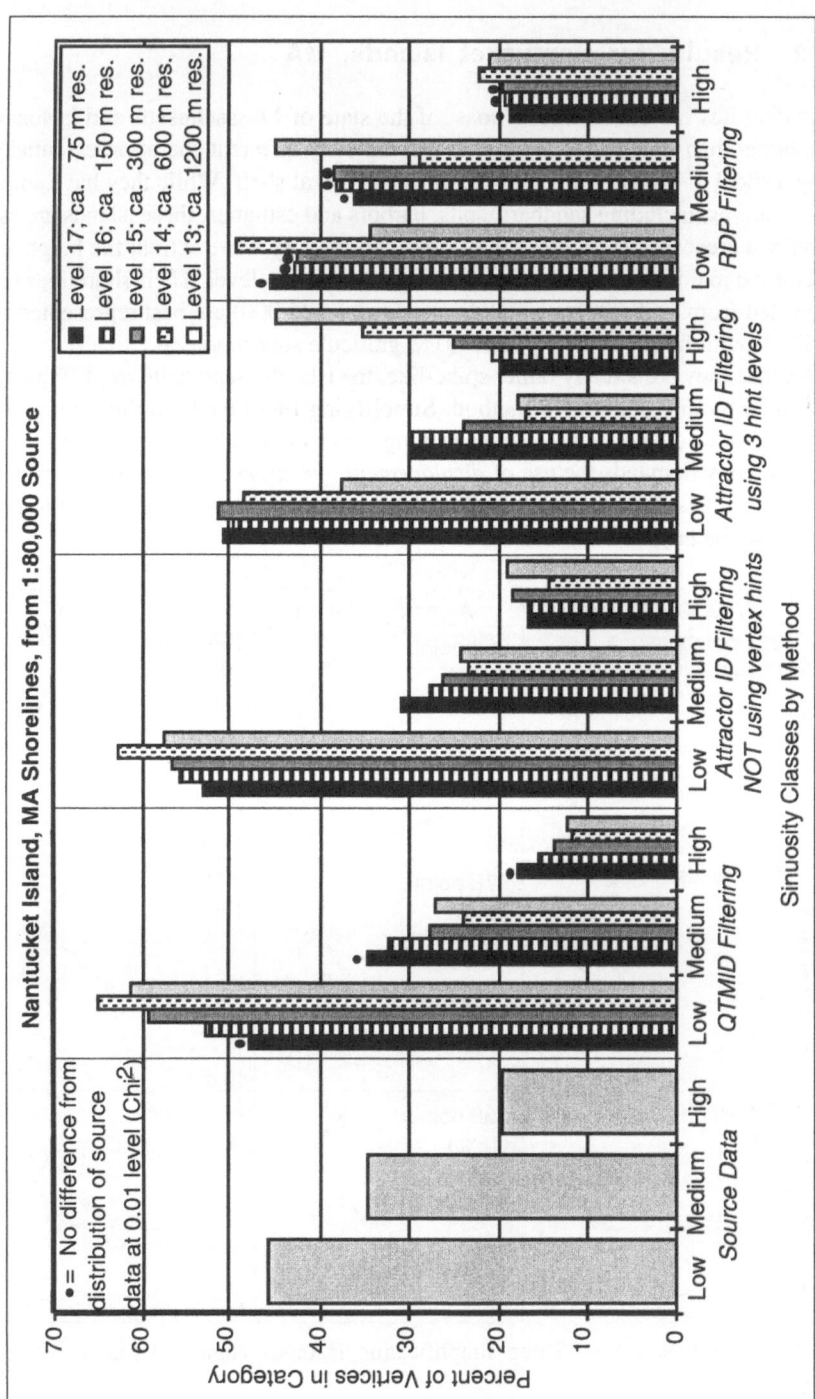

Figure 5.5: Relative distribution of three categories of Vertex Sinuosity before and after applying four generalization methods

An analysis of how different simplification methods affects the shape of the Nantucket dataset was performed in parallel to the one done for Bainbridge shown as fig. 5.3. Nantucket's findings are displayed in fig. 5.5. Both histograms chart relative rather than absolute distributions of sinuosity, partitioned among three classes, and measured across five levels of generalization for four different methods. They show that all methods except RDP create significant shifts in sinuosity, with attractor-based methods changing the distribution more than the quadrant-based method.

When high-sinuosity points are favored (fourth set of bars) the changes are greatest, and become more pronounced as the amount of detail decreases. Had hierarchical generalization been included in this evaluation, it probably would have magnified this effect, as figures C1.2 and C2.2 (based on data mapped in figures D1.11 and D2.11) clearly show. From these limited examples we provisionally conclude that *RDP generalization tend to change line sinuosity characteristics less than QTM methods*. Whether this effect is *desirable* for map generalization remains to be seen. We will return to this issue later on and again in chapter 6.

Lack of Effect of Lookahead. One would expect that the *lookahead* parameter might have had more effect on Nantucket than it did. It should operate to cut back some of the sand-spits by eating away at their tips. This assumes that to get to the tip, the shoreline exits one mesh element (QTM facet or attractor) for another, then returns to the first one along the opposite shore. Whether this happens depends on how features intersect an arbitrary grid, insensitive to shapes. But the major problem most likely lies with the actual values specified for the *lookahead* parameter — which in the experiments shown in appendix D were either 0 or 4. This is probably too small a range to show the anticipated effects, as there are usually more than four shape points describing the end of a spit, and even if there were fewer than four, the resulting truncation would be minimal anyway.

The *lookahead* sub-procedure of *weedQids* is intended to eliminate minor spike along lines, rather than to cause major shape changes, and that seems to be how it works, when it does. Some effects of its use can be seen in figures D1.10 and D2.10, but they are quite subtle and do not appear everywhere they might be desired. Even setting the parameter to a much higher value might not yield the expected result, however, as a spike on a feature may never return to a mesh element visited before. In any event, we have been reluctant to use large *lookahead* values, as they can have the side-effect of eliminating larger shapes than tends to be typical at a given level of detail, generating an uneven degree of simplification along a feature. But this presumed effect is still an open question, and higher extents of *lookahead* should be investigated.

Effects of Sinuosity Selection. Figures D2 through D7 show results of simplifying datasets using *mechanical selection only*; *weedQids* made simply sampled points, making no attempt choose the most appropriate vertex when culling mesh elements. Although the results are certainly not bad, they can be improved upon, as figures D1.11 and D2.11 illustrate. Both test datasets responded strongly to using vertex hint*ing* — sinuosity-based point selection. We find that these effects become more pronounced with greater simplification, which is appropriate. Hierarchical elimination, by causing increased simplification, simply compounds these effects. And in both study areas, we also find that choosing low-sinuosity vertices produces better simplifications than does selecting high-sinuosity ones. Still, not all the consequences are positive ones.

Examine the maps in the top row of fig. D2.11; these versions of Nantucket at QTM level 14 are starting to lose a number of details, especially the map in columns 1 and 3 (low-sinuosity point selection). There are more similarities between columns 1 and 3 and between 2 and 4 than there are between 1 and 2 or 3 and 4, indicating that sinuosity selection more strongly affects line character than does choosing to generalize hierarchically or not. The high-sinuosity maps include much more high-frequency information, which at the intended scales of representation is mostly noise.

There is a difficulty that comes with the smoothness of low-sinuosity points, and that is most apparent on the major sand-spits (if we think of the island as a seated mermaid, these are her neck and outstretched arm). The arm in particular becomes narrower in the low-sinuosity maps, as the high-sinuosity inflection points on the bay side get eliminated. And although her neck tightens a little, the shape of her head is better maintained. What is required is a way to thicken narrow features such as these. As we noted earlier, QTM-based generalization, as implemented, does not provide a *displacement operator* to handle such situations; these particular features, being positive rather than negative space, highlight this lacuna. Our methods handle narrow straits, estuaries and bays better than spits and bars, as eliminating promontories along them tends to open them up, while performing the same operation on a spit tends to close in on its centerline.

5.3 Evaluation and Discussion of Findings

Our experiments demonstrate that using the QTM grid as a spatial filter works reasonably well, at least in terms of the aesthetic quality of map features generalized to date. Even when vertices are selected by mechanical sampling, the character of the features seems to be maintained over a wide range of scales (on the order of 1:32, and possibly more), across which other map simplification methods may not work as gracefully. The parameters we have implemented to control the process provide for many subtle possibilities, only some of which have been examined to date.

Degree of Generalization. As worked examples accumulated that showed the effects of our parameters, we continued to observe that linework was not being simplified by our algorithm to the degree that a cartographer might insist upon. This is evident in most of the maps presented in D2-D7 in the form of small details that would have been eliminated in a manual process. Strategies that seemed to help us most in maximizing the degree of simplification are:

- Using attractors rather than QTM facets as mesh elements,

- Using a Hierarchical Generalization Mode, and

- Selecting only one vertex to represent a mesh element run.

But even applied in concert, these strategies still did not always yield sufficient simplification. We suspect some of these problems are due to having more than one run of vertices occupy a given mesh element; when this happens, even if each run is reduced to a single vertex, there can still be a number of them in remaining in the mesh element. Replacing all the runs with one point would probably achieve sufficient simplification, but this would also tend to introduce topological errors that might be quite difficult to diagnose and correct. So, we must learn to live with the degree of generalization that our current methods provide, and look for better methods to handle specific problem areas. Some ideas that may help in this regard are outlined in section 5.4.

Useful Metrics. Many statistics were collected, derived, analyzed and charted, some of which are reproduced in appendix C. Most of the derived ones are ratios that relate the amount of detail to map scale or resolution. For example, the *SEGMM* statistic describes the average size of line segments resulting from each generalization run at "Standard Scale" (see section 5.2). This value can in turn be used to derive a map scale (and associated QTM detail level) at which *SEGMM* would take on a constant, "optimal" value, such as one millimeter. These scale values are listed in the tables in C1 under the heading *Scale for seg(mm)=1*. Discrepancies between this "ideal" scale and the target scale quantified what we observed above: that our parameter settings — if not the methods themselves — were resulting in too much line detail. The ideal scales could be three to ten times larger than the QTM Standard Scales, which are based purely on mesh resolution, and double at each level.

Figure 5.6, which plots average segment lengths as measured on the ground (SEGKM) and on standard scale maps (SEGMM), shows how variations in methods yield quite different results according to these measures. In the four charts, striped lines depict segment sizes on the ground (in km) and solid lines are the same data, but transformed to the what they would measure (in mm) on maps made at the standard QTM scales (i.e., doubling at each level). Note the cross-tabulated legend: each column presents one of the datasets, using two scales of measurement. Within a row, the units of measure are constant, but different mesh elements are compared.

The erratic behavior toward the far right of the lower series are due to retaining a very small number of points (fewer than 6), at which point the caricatures are no longer useful. But before such a catastrophic scale is reached (above level 12 or 13, around 1:3M), we note a steady rise of ground segment size and slight declines or stability in corresponding map segment sizes. As stated earlier, the most success-ful generalizations seem to yield an average map segment size between 0.5 and 1 mm. The hierarchical mode simplifications tend to produce segments in that range, while non-hierarchical ones produce somewhat less constant lengths that tend to be a bit below the optimum range.

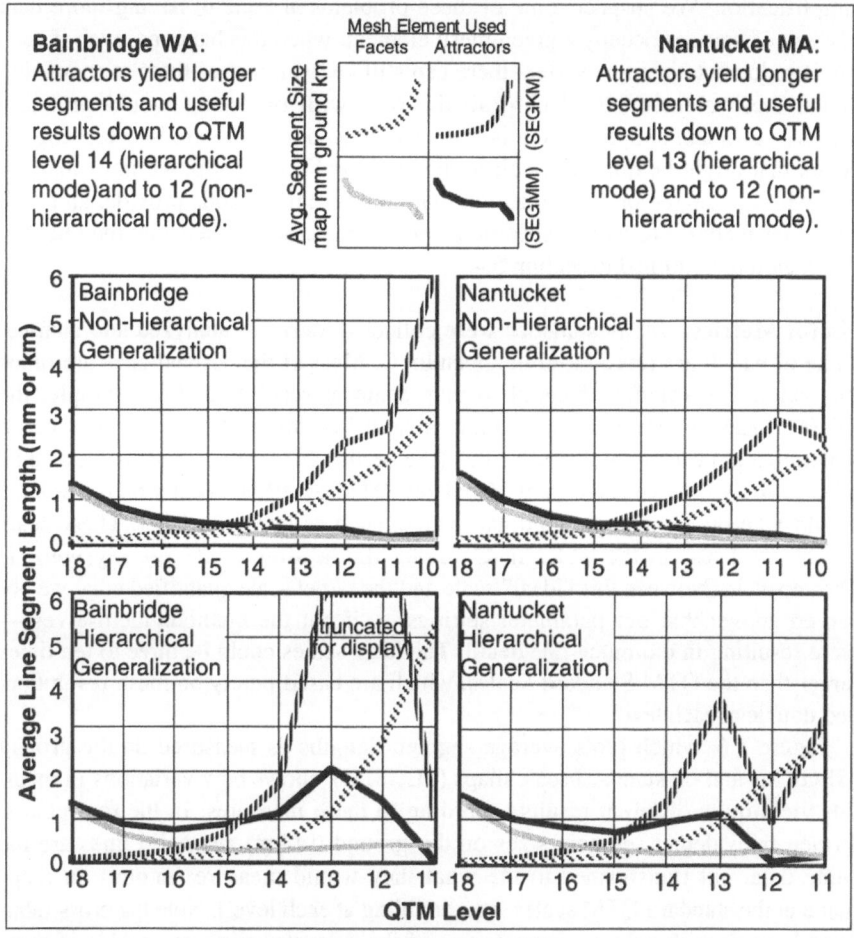

Fig. 5.6: Mean segment size on ground and in map units at standard scales for 8 or 9 levels of detail, by generalization mesh type and mode

Sense from Sinuosity. We found that computing and manipulating the distri-bution of vertex sinuosities can have major and rather interesting effects, and feel more attention should be given to this area. Some suggestions for further research are given in section 5.3.1. The measure of sinuosity we have developed (described in fig. 4.8) is certainly not the only one that could be used, but it seems to behave the way it should in characterizing lines and selecting points. It may be more important how such information is managed and used than how it comes to be computed; as fig. 3.6 illustrates, the QTM encoding scheme makes it possible to build such attribute data into each and every vertex in a dataset, at no additional storage cost (in our character-based implementation, however, it costs one byte per vertex). When vertices are qualified in this or a similar manner (and more than one method can be used in concert), it makes it easier to make decisions about selecting them. These decisions could even be made contextually, based on the purpose of the map, the feature class being simplified and the density of features and vertices in the vicinity (using attractor occupancy information). And segmenting features according to sinuosity — similar to the approach reported by Plazanet et al (1995) — can allow generalization to be chosen more appropriately and their parameters to be more finely tuned.

5.3.1 Comparing Simplification Algorithms

When we compare our methods to the widely-used Ramer-Douglas-Peucker algorithm we find things to like and dislike about each. Results can favor either method or neither. Looking at fig. 5.7, displaying data from Switzerland, it is difficult to decide which method works better for a relatively convoluted feature such as Schaffhausen Canton (many other Swiss cantons have a similar character).

Other comparisons of these methods are shown in figures D1.9 and D2.9, for the two U.S. test areas. In none of these figures was sinuosity selection used for the QTM-based versions, which could well have improved those results. Additional differences exist between QTM and RDP algorithms beside the quality of their results. Before closing this discussion, it may help to compare these characteristics, in order to summarize the difference between our approach and that commonly-used one. This is provided in table 5.2, and further elucidated below.

	QTM	RDP
Operational Scope	local	global
Hierarchy	Y/N	Y/N
Parameters	6-8	1
Uniformity	higher	lower
Useful Scale Range	ca. 1:32	ca. 1:8
Point attributes	used	not used
Span of Control	large	small
Comprehensibility	complex	simple

Table 5.2: Characteristics of Simplification Algorithms Compared

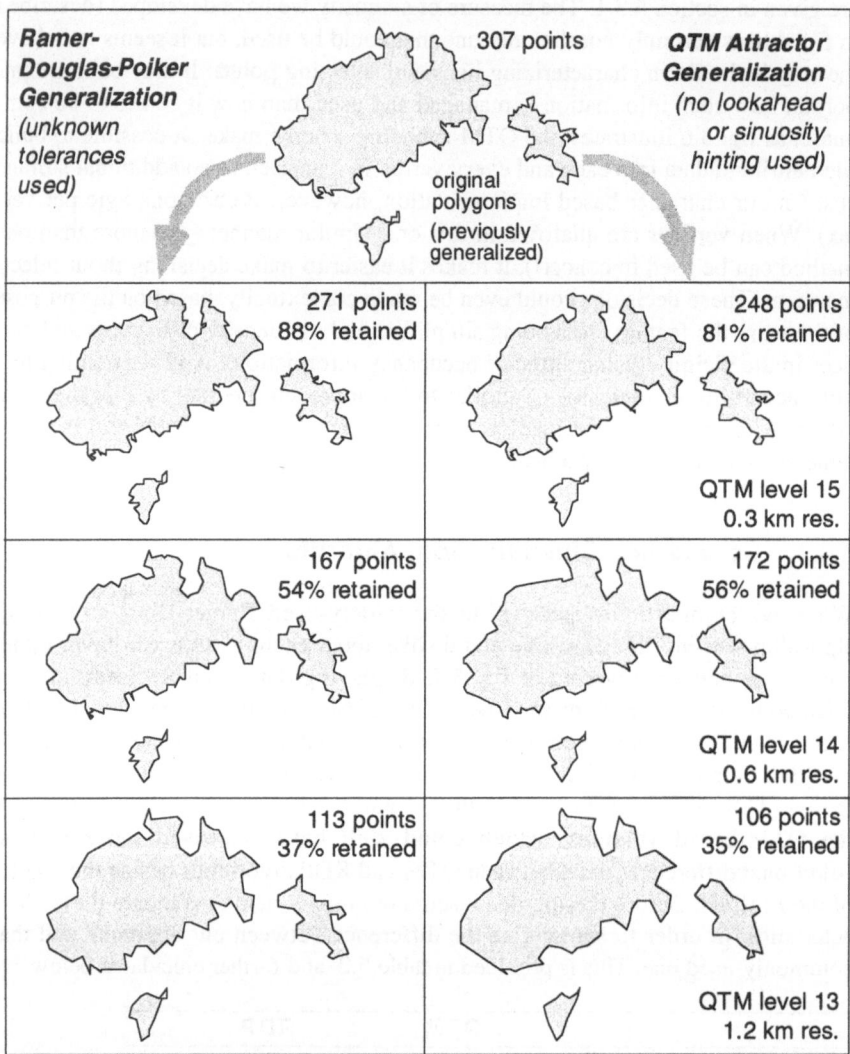

Fig. 5.7: Schaffhausen Canton, Switzerland: Comparison of QTM and RDP generalizations having roughly equal numbers of points.

<u>Operational Scope</u>. While QTM is a global coordinate system, as applied to map generalization its operators work locally, usually looking only at the contents of one mesh element at a time. RDP is a global operator, in the sense that it treats each given feature as a unit of analysis. Its results can be biased by the choice of these units (it is sensitive to where arcs and polygons begin and end). QTM-based methods are sensitive to the placement of features with respect to mesh elements.

hierarchy. Both approaches can be implemented hierarchically, but this is not a defining characteristic of either one. By modifying their logic, one can ensure that vertices will be selected top-down, so that points eliminated at small tolerance levels will not suddenly reappear at coarser levels of detail. By working this way, features can be pre-processed for rapid display at a later time. Barber et al (1995) studied the effects of hierarchical elimination in great detail, and concluded that using it caused few perceptible or statistical differences in comparison to non-hierarchical algorithms.

Parameters. Section 4.4.3 outlined eight parameters or modes that affect QTM generalization, and most of these have interactions of some sort. RDP has but one parameter, which like some of QTM's, is not very intuitive. Setting a bandwidth tolerance properly can be frustrating and take time, whereas specifying a QTM level of detail is more easily grasped (but other parameters are less obvious).

Uniformity. Despite not being a global procedure — or perhaps because of it — QTM simplification tends to yield rather uniform results along primitives, regularizing vertex spacing in the process. When attractors are used as mesh elements, two types of interlaced faces result (triangular and hexagonal cells), but their effects blend together imperceptibly. RDP, on the other hand, neither respects nor coerces uniformity, and can cause point density to vary erratically along features. Such non-uniformity is unpredictable, depending on the choice of anchor points, tolerance levels and local geometry. The most common results are featureless spikes and long, straight lines lacking character, where intermediate details get swallowed by the band imputed to lie between two anchor points.

Useful Scale Range. Problems can occur when RDP is used to greatly change the scale of representation of map displays. Often these take the form of self-crossing features in areas of complex geometry. Topological post-processing normally is invoked to eliminate such overlaps. QTM simplification is not immune from these artifacts, but they are relatively rare, especially when points of low sinuosity are selected. This indicates that "characteristic points" such as RDP identifies may not always be the best choice. This notion is explored in the next chapter.

Point attributes. Because it is possible to coerce the *weedQids* algorithm to select certain vertices in preference to others, users potentially have some control over results. RDP could be modified to work in a similar way (for example, by declaring certain vertices to be "sacred"), but this would essentially create a new algorithm, having unpredictable properties. Also, QTM's built-in vertex attributes (metadata) need not be limited to measures of local sinuosity, as is now the case.

Span of Control. By availing themselves to multiple parameters, QTM-based methods are inherently more flexible than RDP, or any algorithm with a single parameter. If special generalization situations can be identified, better results can probably be obtained by tinkering with one parameter or another. We are only just now learning how to do this. The experiments with segmenting features according to sinuosity are a start in this direction.

Comprehensibility. One usually pays for flexibility of control, and the price is often confusion and potential for making mistakes. This is the dark side of span-of control, its evil twin. It will take some time before QTM generalization is well-enough studied to understand its behavior. If it is going to succeed, some degree of automated parameter setting will be needed in order to routinize its application, such that GIS users can call on it without being constantly confused. While they need not necessarily understand how QTM-based algorithms work in all their detail, they will need more help than their software or data normally provides them.

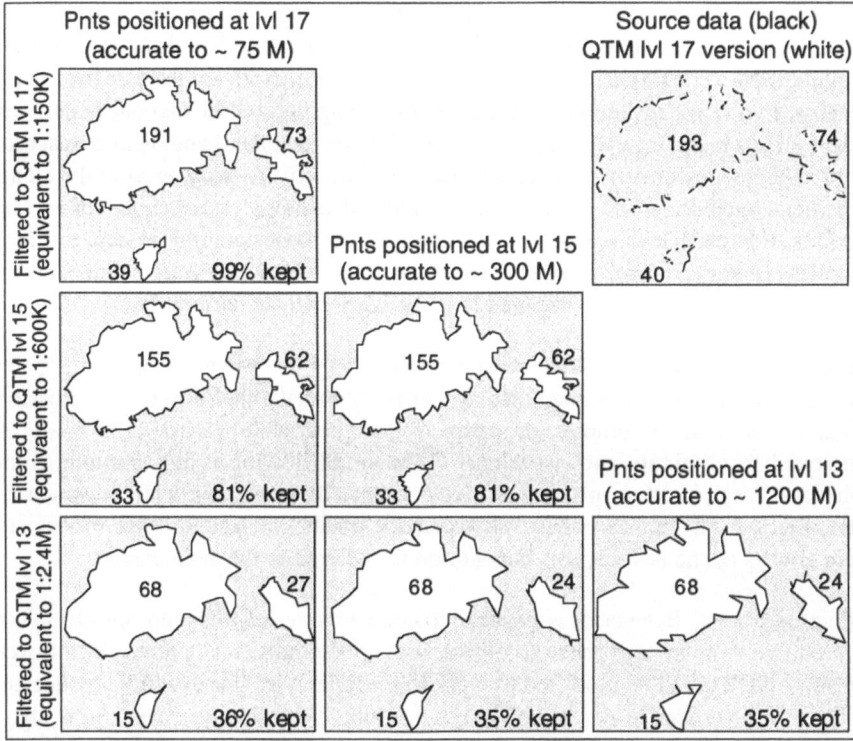

Fig. 5.8: Effects of Varying QTM Retrieval and Point Decoding Resolutions. The QTM resolution used to *select points* (detail) varies across rows; QTM resolution used to *regenerate coordinates* (coarseness) varies across columns

Point Placement Possibilities. There are additional ways in which QTM generalization results can be subtly modified. Figure 5.8, using the Schaffhausen boundaries, shows some of these other possibilities; it controls two QTM parameters, QTM *retrieval depth* (by now familiar to readers), and one not usually varied, QTM *decoding depth* (discussed briefly in section 4.4.2). This determines the precision with which QTM IDs are translated back to latitude and longitude. Normally full precision of vertices (*endoding depth*) is used, but in certain circumstances it may help to coarsen the locational precision of features, particularly for choropleth thematic maps.

At extremely low levels of decoding precision (as shown in the lower right of fig. 5.8) spikes can appear in odd places. These result from aliasing the base vertices of small features such as peninsulas which reduce to triangles. In such situations one might count on the *lookahead* parameter to delete the spikes, but only if it is set high enough to traverse all the vertices from one base point to another lying in the same mesh element. A better way to handle this might be to *post-process filtered vertices* to identify spikes at retrieval resolution rather than at source resolution. If these minor artifacts can be dealt with by this or other means, the *decoding depth* parameter can probably be made more useful.

5.4 Chapter 5 Summary

This chapter presented and evaluated findings of empirical tests of line simplification methods described in chapter 4 (and formally presented in appendix A). It began by describing the computing environment used to create a prototype map data processor called *QThiMer*, written in the C programming language. Other software used in conducting tests was identified, and the general procedures used to prepare graphical and statistical summaries of results were outlined. Criteria for selecting the test data were discussed: it was desired to use natural map features of different geomorphological types that were complete in themselves rather than being portions of larger features; small enough to be displayed at large-to-medium scales on a normal page; containing a variety of types of details. These criteria seemed best met by islands, ones smaller than about 20 km; good examples were then found in the U.S., offshore of Massachusetts (Nantucket) and Washington States (Bainbridge), and the coordinates of their shorelines were downloaded from a WWW site operated by the U.S. Geological Survey.

Results from generalizing the shorelines of each test site were then discussed separately. Several methods and parameters were applied identically to both datasets, and the results were presented in tabular and graphic forms to facilitate comparisons. Results were evaluated with respect to the Ramer-Douglas-Peucker (RDP) line simplification algorithm, the dominant (and sometimes the only) generalization method found in GIS packages. It was found that the two approaches often yield similar simplifications (especially where reduction ratios are low), but certain important distinctions were noted. This led to a discussion of the modeling of line sinuosity, and the use of sinuosity statistics in both performing and evaluating line generalization. One notable finding was that selecting map vertices to minimize sinuosity always seemed to generate better simplifications than selecting vertices to maximize that statistic. In concluding discussions, additional findings concerning effects of parameter settings were described, in an effort to characterize line simplification methods. The concept of sinuosity and its operationalization merits further exploration, and refinements in both our methods and our thinking were suggested by the experiments reported above. Some of these were formalized, implemented and tested, and this work is reported in chapter 6.

6 New Approaches to Cartographic Questions

Things derive their being and nature by
mutual dependence and are nothing in themselves.
— Nagarjuna

Among other insights, this project has given us a new perspective on what cartographers call "characteristic points", describing major inflections along a feature that contribute more overall shape information than most points do. Prior to our experiments, we assumed that the locations of such points were knowable only after analyzing an entire feature as a whole. The notion comes from the work of psychologist Attneave (1954), and into cartography via Peucker (1975), Buttenfield (1985) and others, including computer vision researchers (Ballard and Brown 1982). In this chapter we will present a different approach to thinking about feature characterization and point selection, after describing the context in which line simplification techniques have developed and continue to be evaluated.

6.1 On Cartographic Lines and Their Simplification

On analog maps, lines usually have beginnings, ends, widths and flow from place to place as continua, with occasional corners. In digital maps, lines have beginnings and ends but not widths, and are sequences of dimensionless points, each of which may have little intrinsic importance, often representing nothing in particular. Digital cartography has, however, instilled in its practitioners a sense that such points really do exist, and that their roles and significance can be quantified and classified. Perhaps that is why—seen from other perspectives—the literature on map generalization often seems to make much ado about nothing, or almost nothing. In this section, we attempt to sort through some of the controversy surrounding line simplification, to see what is really at issue and if perhaps some deep assumptions exist that may be—at least sometimes—counter-productive.

6.1.1 Imperatives of Scale

Generalization is not simply making little things look like big things. Children are proportioned differently than adults, and map features at small scales should not slavishly mimic their shapes at large scales. The purpose of the display or map should also influence shape representation, as section 6.1.6 describes. We should also make clear that simplifying features may require more than simply removing vertices; sometimes entire shapes (such as a meander or a switchback) should be deleted at a certain scale. In other instances entire features (arcs, polygons or sets of them) will need to be eliminated, or their symbolism class may have to change. But long before that happens, a feature will tend to lose much of its character as a consequence of being reduced. This needs to take place in a sensible way, and as the following sections describe, commonly-used numerical criteria for this process do not necessarily lead to satisfactory solutions.

6.1.2   Weeding and Conditioning Data

Much early work in generalization, especially that deriving from Jenks (1981) (also see McMaster 1987) focused on conditioning raw digitizer data and reducing their volume to minimize storage requirements, at a time when most people digitized their own data and disks and memory were quite expensive. This was also the motivation behind Douglas and Peucker's (1973) approach to point reduction. Now map data are often taken off the shelf (e.g., read from CD-ROM or internet sources), and are presumed to be already conditioned by their vendors (but of course errors and inconsistencies continue to be inevitable). There may even be a choice of scales for commonly-used layers. This means there are different kinds of generalization problems, and not all the old concerns are still salient. But a surprising number still are, and common procedures used in interactive mapping (such as zooming or viewing arbitrary combinations of layers simultaneously) add even more problems to the list.

6.1.3   Local versus Global Strategies

Along with other characteristics, line generalization algorithms are described as local, regional or global in scope (Douglas and Peucker 1973, Zycor 1984). Aside from the seldom-seen Fourier method (Moellering and Rayner 1982), there is really only one global algorithm in use, RDP (Ramer-Douglas-Peucker; see chapter 5). It dominates in part because of that characteristic, which almost everyone assumes is helpful. Starting with Attneave (1954, who studied drawings rather than maps), the idea that certain locations along lines provide greater information about line shape than others do became part of cartographic conventional wisdom. Because RDP is superficially good at finding such locations, it was assumed it should offer a sound perceptual basis for line generalization, and this was reinforced by a number of studies (Marino 1978, White 1985, Buttenfield 1984, Jasinski 1990). However, other studies have shown that RDP can fail to yield generalizations that are visually acceptable (Visvalingam and Whyatt 1990, Beard 1991, Zhan and Buttenfield 1996). Jasinski's study might also have shown graphic shortcomings of RDP, but did not present enough plotted examples for readers to determine this. Over the years the RDP algorithm became implemented on many GIS platforms, and its ubiquity along with the findings that it produced recognizable simplifications and good metrics, seem to have blinded many cartographers to the fact that it often produces bad-looking lines.

It is also widely acknowledged that RDP is sensitive to initial conditions; in commercial GIS these tend to be dictated by line topology and are not often (or easily) changed. However, while some work has been done that describes consequences of changing initial RDP anchor points (Buttenfield 1986, Visvalingam and Whyatt 1990) little experience and few practical guidelines exist in this area. New approaches to feature segmentation, including work by Plazanet et al. (1995) and the investigations reported in chapters 4 and 5 are going after such problems, not necessarily using RDP. We suspect that by segmenting lines to be more homogenous, then applying appropriate algorithms and parameters to each regime individually will almost always improve results. To do this with RDP would require re-setting the tolerance parameter rather frequently, which could burden users with difficult-to-make decisions, unless algorithms were developed to do this.

6.1.4   Choosing Tolerance Values

Tolerance-setting for line generalization is normally left to users to decide on a case-by-case basis, and very few guidelines or automated assistance are generally available for this. Some simple algorithms have tolerances that are fairly intuitive, but this is not the case for RDP. As Cromley and Campbell (1992) discuss, there may be no simple relationship between tolerance values and intended display scales, as the effects of changing the parameter may be non-linear, making the process difficult to tune. They and others have experimented with determining the number of points to be retained by applying the radical law (Topfer and Pillewizer 1966) to the scale reduction factor, using a modified RDP that will simplify to provide the number of points required. A study by Barber et al. (1995) indicated that doing this yields very good statistical results when commonly-used measures such as vector- and area displacement are made of the resulting linework. Even when tolerances are subsumed by a number-of-points criterion, and thus become less abstract, poor settings can still result in self-crossings, uneven densities and other unwanted artifacts. Some of these problems can be due to the given initial endpoints, which as mentioned above are rarely modified. Others are simply direct consequences of choosing the most salient points, as we will further discuss.

6.1.5   Optimizing Performance

There is more to good simplification than just cutting down clutter to acceptable levels. Cromley and Campbell (1992) demonstrate quite clearly how success also depends on *which*, not just on *how many* points are selected by applying linear programming techniques to enforce certain geometric criteria for point selection. They compare this type of optimization to RDP using a digitization of the coast of Connecticut, minimizing vector displacement, and both minimizing and maximizing line length. RDP is shown to produce lines ca. 90% as long as the latter optimization, but of very similar visual quality. The minimum-length optimization solutions are extremely smooth, having lost almost all hint of the bays and estuaries characterizing this coast, and the authors describe them simply as "bad". That, however, is a premature judgement which takes no account of the intended scale and use of the simplified data.

6.1.6   Assessing Performance

Now-standard cartometric measures such as total length, segment size, vector displacement, areal displacement, angularity and others have objective validity and have often been used to quantitatively compare generalization solutions (McMaster 1983, 1986; Jasinski 1990, Barber et al. 1995). However, if used as guides for selecting procedures or parameters they can be quite misleading. This can happen when one's goal is to produce caricatures that most closely resemble input data, to the exclusion of aesthetic concerns such as were described in section 6.1.1, and trying to satisfy global criteria rather than to optimize within local contexts, as discussed in section 6.1.3. McMaster (1987), following Jenks (1981), addressed this general issue, attempting to typify goals of line simplification via a 3-dimensional framework having continuous or quasi-continuous orthogonal axes. This took into account 1) output scale, 2) output resolution and 3) map purpose, but the goal still remained "to select the salient features along a digitized line" (McMaster 1987: 100).

However, there is some controversy if not confusion about how "salient features along a digitized line" might be identified. The consensus seems to be that these are often the same critical or characteristic points identified by RDP. Nevertheless, there is disagreement, and some evidence to the contrary. Mentioned above, Visvalingam and Whyatt (1990) conducted a detailed evaluation of RDP's performance, comparing it to manually generalized versions of the source data (British coastlines). A variety of differences were noted, almost all favoring manual generalizations. The authors note "Points selected by the Douglas-Peucker algorithm are not always critical. Manual generalizations take into account the relative importance of features. This is partly dependent upon the purpose of the map. Even if we ignore such variable factors, manual generalisations tend to preserve the shape of geometrically large features at the expense of smaller ones" (Visvalingam and Whyatt 1990: 224).

6.2   Important Compared to What?

We concur that while map scale, output resolution and purpose should be always borne in mind, selection of "important" characteristic points, in being so blindly emphasized, causes simplifications to be sub-optimal in many cases because the most descriptive shape points at one scale may be less so at another, and therefore may be bad choices; "just because the cat had her kittens in the oven don't make 'em biscuits", a New Englander would say.

In the experiments reported in chapter 5 we sometimes noticed that *points that would not be regarded as globally characteristic may be better candidates for selection during generalization*. Even though such points are "less important", than more obvious inflections, they nevertheless can be more representative of local line conditions; the nature of their localities rather than their global *gestalt* determines their cartographic utility. We will call these (variable) extents of lines *regions*, to distinguish them from truly local neighborhoods (such as a point and its immediate one or two neighbors).

6.2.1.   Generating Good Gestalts

If, as we found, choosing less significant points can generate superior generalizations, and if, as we and other researchers postulate, global line simplification methods such as RDP sub-optimize by failing to identify appropriate contextual neighborhoods, what can be done to improve their results? While it is clear that RDP can handle the latter issue to some extent if arcs are segmented into more natural units, addressing the former one necessitates a completely different approach to what constitutes characteristic points. The concepts, machinery and metrics of QTM-based line simplification, as outlined in the prior two chapters, can be used to implement alternative methods that might work better, as this section will describe.

Regional Gestalts. The data model presented in chapter 4 enables each vertex along a primitive to use left-over space in its QTM ID to record measures of sinuosity. Such "enhanced identifiers" were implemented by assigning a certain byte to represent vertex sinuosity. Should a primitive have no sinuosity data, the byte will contain zero. Two types of sinuosity indices have been described that use this field, local values or more global estimates. However, both types of sinuosity estimates could be computed and stored, space permitting. As no coordinate data in this study requires more than 20 QTM digits to describe, lack of space is not a problem. In our implementation, the first byte can store a sinuosity index, (scaled from 1 to a maximum of 15) for that specific vertex, computed in a way that *emphasizes the role of the vertex in its immediate locality*, usually three or fewer points on either side of it. The second byte contains a similar index (computed using the same algorithm and perhaps having the same upper limit), but it *characterizes the sinuosity measured across a larger neighborhood and smoothed if necessary*, as figures 4.8 and 4.10 describe. For most primitives, the first (local) index will tend to have greater variability than the second (regional) one, because it is less well damped by other values in its vicinity.

These two statistics may be used together to select points in a context-sensitive manner. When generalizing, and no user-defined preferred sinuosity value (PSV) has been declared, the *weedQids* algorithm can use the regional (second) byte in that role, to select the point in a run that has a local sinuosity value (first byte) closest to it. When there is more than one vertex in the run that match according to this criterion, the one closest to the middle of the run will be chosen in preference to the others (i.e., using the median criterion), as this tends to space vertices a bit more evenly.

The result of doing this should be *to select points that are most typical of their neighborhoods*, based on whatever local line complexity metric is employed. This should help to *preserve line character* throughout the primitive, which one might presume to be theoretically desirable and a primary goal of simplification.

But is it? One of the problems of line simplification algorithms we have identified is that complex portions of lines tend to be turned into tangles, or at least awkwardly crooked shapes, while all the character is squeezed out of gentler contours. We suspect that selecting globally — and possibly even regionally — characteristic points does not maintain local line character, rather this tends to *exaggerate* it. If one regards high-sinuosity points as characteristic ones (because they tend to stick out from their local neighborhoods), why does selecting them not provide superior simplifications?

The answer may be hidden in the question: in environs where most vertices are not sharp or highly distinct, those that are may not be truly representative. In any locality, choosing high-sinuosity vertices is tantamount to asking for more jaggedness; this might be appropriate in some areas, but obviously not everywhere. Likewise, selecting low-sinuosity vertices in areas of low line relief will tend to quiet such areas down even more.

A Counter-intuitive Approach. It is thus possible that selecting vertices to maintain the general sinuosity of their localities — as described above — would result in the same type of exaggeration as other methods generate, to the detriment of the graphic result. Yet it might be quite easy to prevent this from happening: simply *select vertices whose sinuosity is least like that of their neighborhoods*, rather than most like them, as above. This should help to tame details in complicated regimes while better maintaining inflections in simpler ones. In regimes of medium sinuosity, however, this rule could decide in either direction, less predictably and thus less consistently. In such cases a user-specified PSV might be used to choose vertices. Wisely chosen or not, this would at least enforce some degree of consistency.

Based on *weedQids*, described and demonstrated in prior chapters, a more refined point selection algorithm has been developed that evaluates vertices by computing a *selection score*, the absolute difference between its local and global sinuosity values. Within a run, those vertices with the highest selection score are preferred. If there is more than one vertex with the highest score, the user's PSV is used in place of the global value to select one or more of them, but this time choosing vertices having the minimum absolute difference. In the absence of a PSV, the default is to select the least sinuous vertices. If, after applying PSV there are still several candidate vertices, the one closest to the mechanically-selected median (and the prior rather than latter, if there are two) is then chosen.

The user's PSV thus is invoked only when the sinuosity selector identifies more than one candidate vertex in a run. Ties are broken in favor of vertices most similar to PSV, and should this set contain more than one member, a winner is chosen using a topological criterion. The method, like ones presented in previous chapters, can be invoked in either non-hierarchical or hierarchical mode.

To recapitulate this point selection approach, a nested set of decisions is triggered each time a run of vertices in a mesh element is identified:

0.   If the run contains only one vertex, select it.
1a. Select points having the minimum absolute difference between
        CSL and CSR or
1b. Select points having the maximum absolute difference between
        CSL and CSR
2.   Select points having a CSL closest to a user-specified sinuosity
3.   Select from whatever candidate points remain the one nearest
        the median point.

Selection always halts at any stage where only one candidate vertex remains. Either step 1a or 1b or neither is taken, followed optionally by step 2, and ending with step 3 whenever more than one point remains to be eliminated from a run. Should only one candidate point per run remain after any step, subsequent steps will not be executed. The difference between 1a and 1b is that in 1a points having similarities typical of their regions are selected, while in 1b, points most different in sinuosity from their regions are chosen. In step 2, users specify a CSL value (PSV) that selection should prefer, should choices remain; any value within the number of sinuosity categories can be given, but in tests illustrated below only extreme valuess (1 or 7) were used. The algorithm requires from one to five passes through each run, depending on which steps are executed; steps 1 and 2 each need two passes; first to find the value to prefer, next to select the point(s) possessing it. However, all computations are simple integer or logical operations.

We wish to underscore the generality of the above algorithm. Even though the procedures described above use QTM-encoded points, *the principles our methods use could be adapted to generalizing points encoded in plane coordinates.* QTM IDs provide a convenient way to transport the sinuosity data without referring to external lists, and the QTM grid provides a sampling framework for initial point selection. But other forms of point sampling could be substituted and the method would work the same, although geometric outcomes would no doubt differ. This is because the procedure makes no reference either to QTM IDs or coordinates, only to attributes of a sampled set of points. Finally, while we propose to use our sinuosity measures as point attributes, any other estimates of local line complexity could be used, even RDP distances-to-trendline or rankings of them.

6.2.2   Comparing Regional and Local Sinuosity Selection

Figures 6.1–6.4 are all arranged similarly, and all illustrate effects of several parameters when generalizing coastline data to a target scale of roughly 1:2.4M. The larger maps are shown at 1:350K (Bainbridge) and 1:500K (Nantucket). Figures 6.1 and 6.3 show generalizations of 1:80K source data, while figs. 6.2 and 6.4 use the 1:250K source data. Within each figure, the source version is shown in the upper left, and next to it is the default QTM generalization (obtained by sampling median points of runs without consulting the computed point sinuosity attributes). To the right of that is the RDP generalization and its tolerance, (set to produce the same number of points as the QTM filtering yielded). Note that the number of generalized points for 1:250K sources is fairly close to the number of points derived for 1:80K data. This is to be expected, due to selection through a fixed grid, assuming both source versions are properly geopositioned and are not too dissimilar to start with. The average sinuosity of each version is printed below it (on a scale of 1 to 7). Each generalized version is reproduced as an inset at target scale to the lower left of the enlarged feature. The insets are essential for interpreting results of methods, as reproduction to scale is usually the primary motivation for generalization, and results need to be legible at target scales. To aid interpretation, boundaries of the enlarged versions are thickened to reflect the line weight of the target scale figures (0.4 mm) as if photographically enlarged to the larger presentation scales

The lower two rows of generalized features are variations of QTM filtering that vary several sinuosity-related parameters. Vertices of features in the middle row were selected that have classified local sinuosity values (CSL) closest to their regional ones (CSR), *so that selected points are representative of their localities.* Conversely, the bottom row contains features having selected points with CSL farthest from CSR, so that in that row vertex sinuosities tend to be unrepresentative of their localities. Such differences are not always visually apparent, especially at larger scales where there are fewer choices available, because *mels* (hence runs) are smaller. The Bainbridge maps (figs. 6.1 and 6.2) show these differences better than Nantucket maps do, because most of Nantucket's most expressive details are already filtered out at this resolution Note the extent of QTM facets at level 13 is.ca. 1.2 km; the hexagon at the top left of each figure shows these *mels* at at 1:350K (Bainbridge) and 1:500K (Nantucket).

The three columns across the two bottom rows show the effects of specifying a sinuosity value for breaking ties between vertices once the minimum/maximum similarity criterion had been applied. That is, having identified all the points in a run whose CSL was most dis/similar to their CSR, the user then can select those which have a CSL closest to a particular preferred value (PSV). Should a PSV not be specified (column 1), median selection mechanically picks the point nearest the middle of the run. If a low value of PSV is specified (column 2), the resulting outline ought to be smoother, at least in the middle row. In the lower row, this logic tends to reverse, as the set of remaining points may tend to mitigate the effects of PSV. In the third column, specifying a large PSV usually results in higher angularity, and the accompanying sinuosity statistics tend to bear this out. Note that the average sinuosity is computed using grouped data (with seven bins), and that it represents the sinuosities of the points selected from the source data, rather than a re-computation for the generalized versions; were sinuosities to be re-computed from the selected points, they would tend to increase, but still maintain their relative magnitudes. Such increases are due to the local averaging process, the range of which extends farther as the number of coordinates decreases, and (in the case of RDP) because higher-sinuosity points tend to be retained. QTM filtering, on the other hand, may result in lowering sinuosity. The graphic results indicate this can be desirable, as it tends to produce simplifications that work better at the target scale, even though at enlarged scale they may appear too smooth.

Although interpretations can vary, results from the experiments suggest that points which would not be regarded as globally salient may be better candidates for selection during generalization. Even though such points are "less important" than more obvious inflections, they nevertheless can be more representative of local line conditions; the nature of their localities rather than their global Gestalt determines their cartographic utility. In addition, regions where shape complexity is above average (such as Nantucket Harbor) may require different treatment to be properly generalized. Feature segmentation techniques (such as investigated by Plazanet et al. 1995; our approach is illustrated in appendix figures D1.12 and D2.12) can help to handle such situations. Finally, our findings underscore the importance of visually evaluating any generalization of spatial data at its target scale (and symbolism), not just in comparison to source data. We expect shapes of features to change as they simplify; this is the inevitable consequences of scale change, unless one is viewing a mathematical abstraction such as a fractal curve. And even when geographical features happen to be self-similar, this does not in itself require cartographic generalizations of them to resemble them in all respects at all scales. Instead, like melting ice sculptures or fragments of holograms, map features should gracefully degrade as they shrink, their importance draining away into their context.

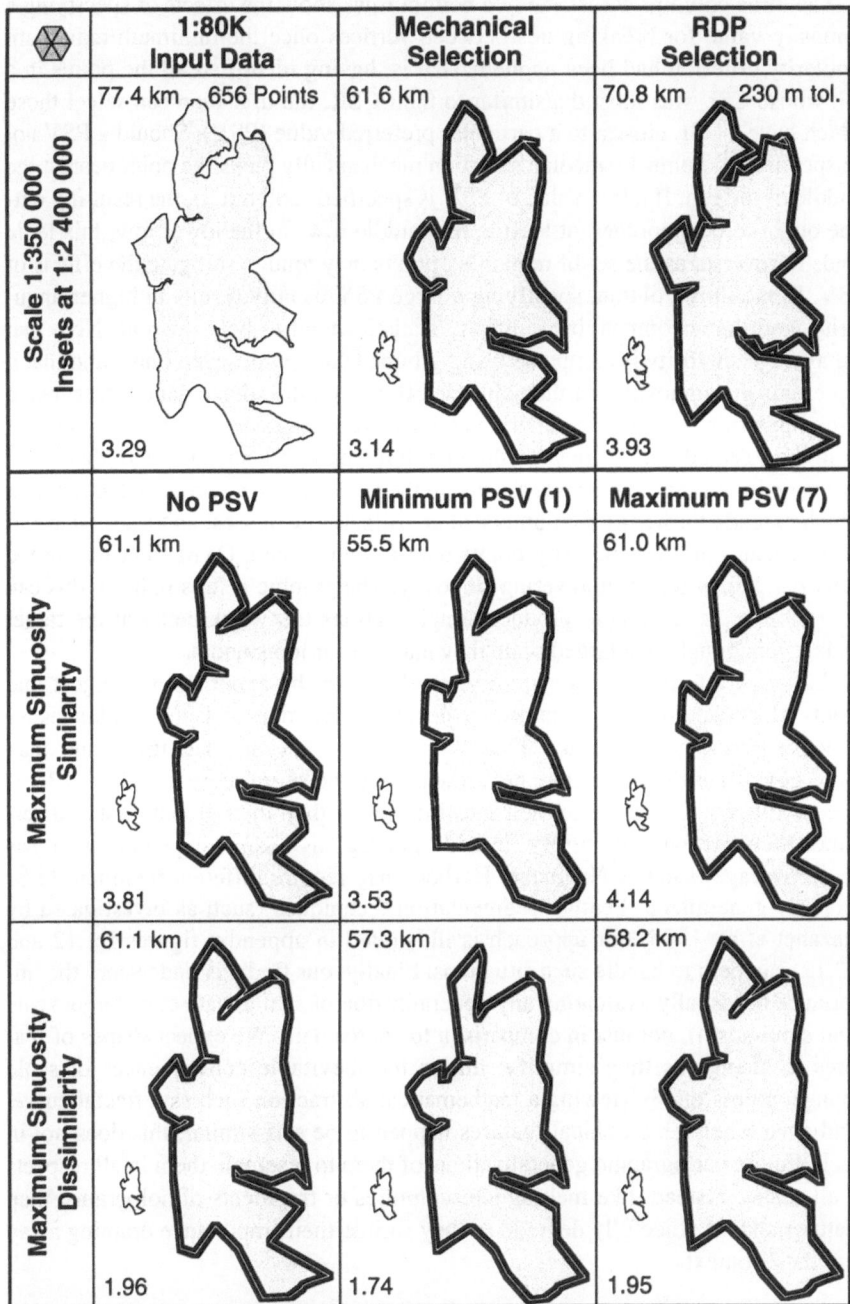

Fig. 6.1: Bainbridge Island, 1:80K source generalized to QTM level 13 (ca. 1:2.4M). All generalized figures have 57 vertices (9%).Sinuosities beneath figures are scaled from 1 to 7. Coastline length is shown at upper left. Line weight multiplied by 7.

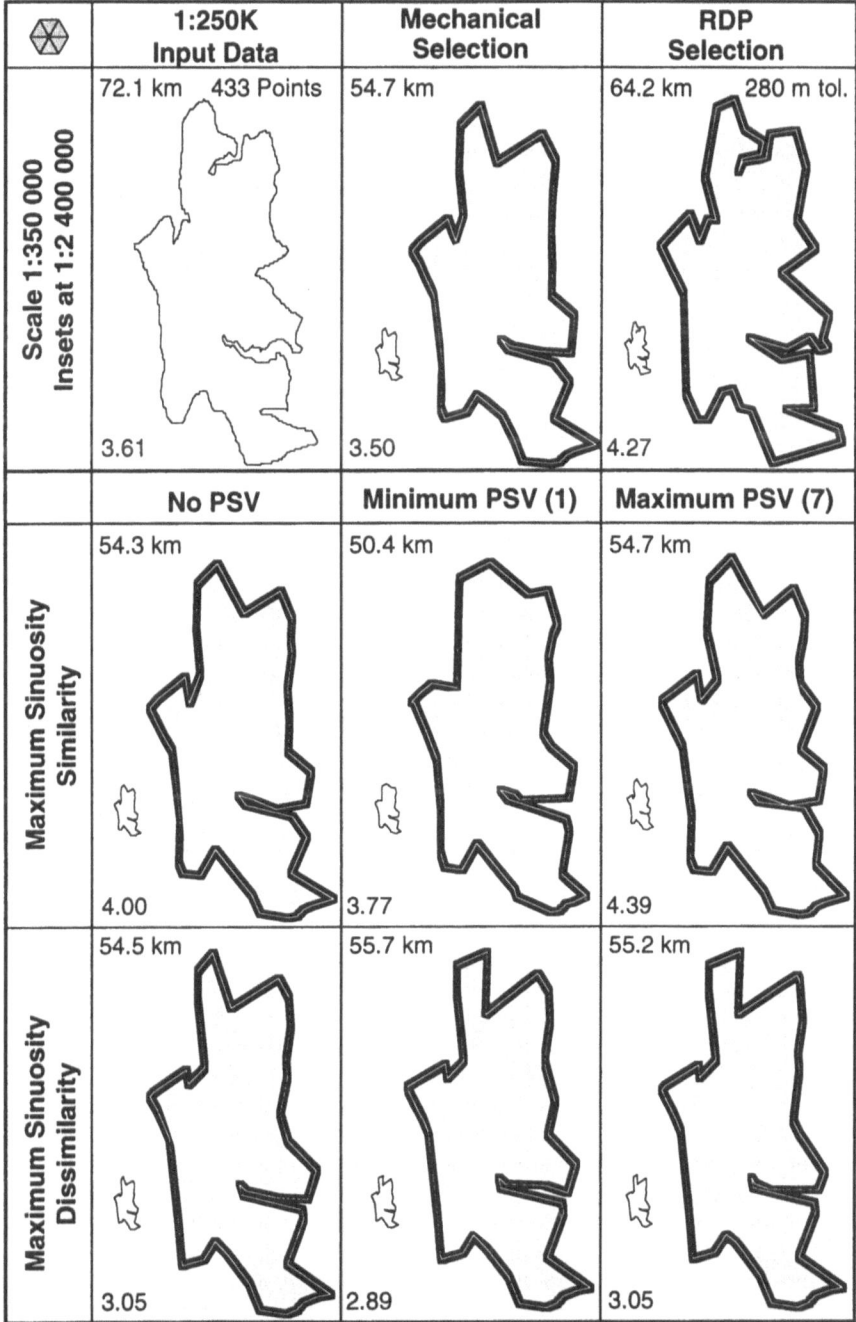

Fig. 6.2: Bainbridge Island, 1:250K source generalized to QTM level 13 (ca. 1:2.4M). All generalized figures have 44 vertices (10%).Sinuosities beneath figures are scaled from 1 to 7. Coastline length is shown at upper left. Line weight multiplied by 7.

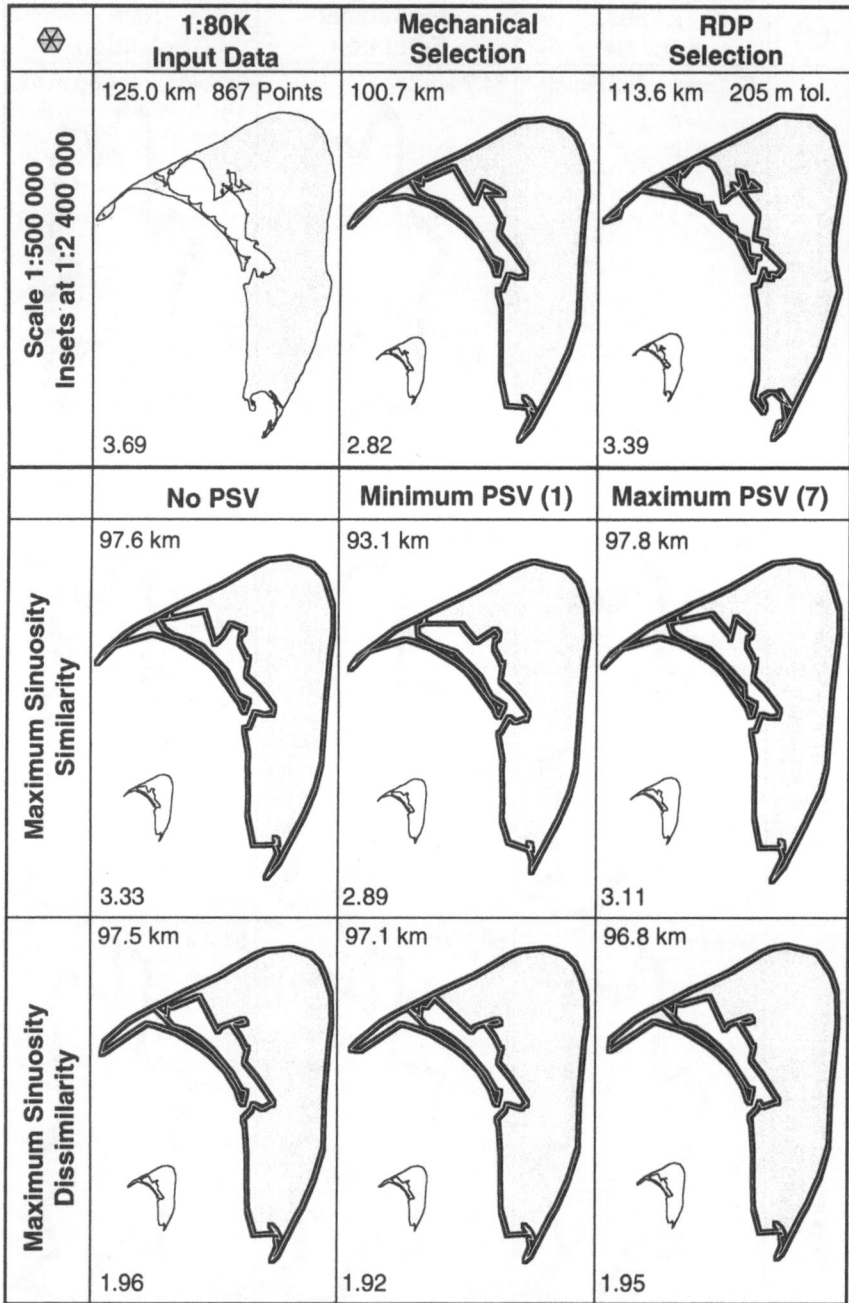

Fig. 6.3: Nantucket Island, 1:80K source generalized to QTM level 13 (ca. 1:2.4M). All generalized figures have 585 vertices (10%).Sinuosities beneath figures are scaled from 1 to 7. Coastline length is shown at upper left. Line weight multiplied by 5.

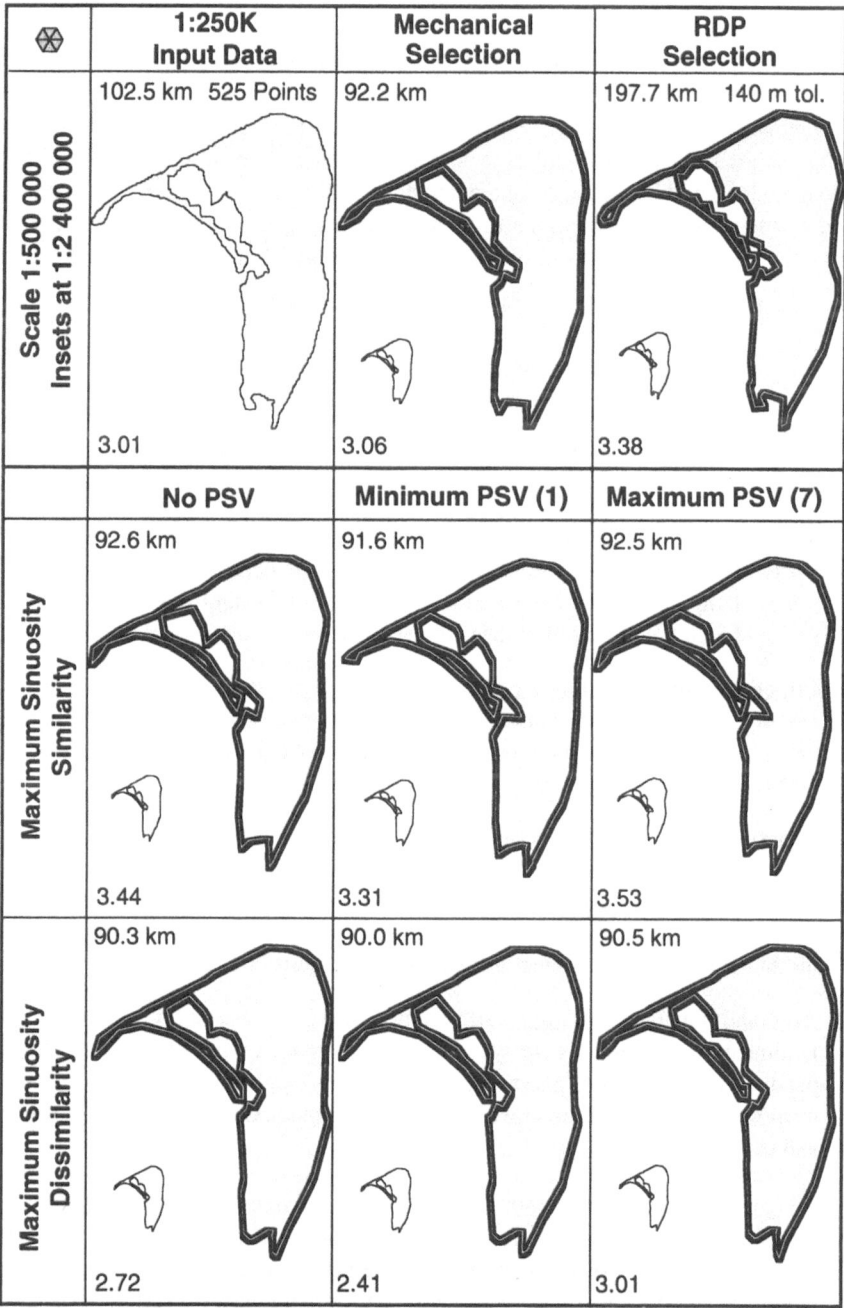

Fig. 6.4: Nantucket Island, 1:250K source generalized to QTM level 13 (ca. 1:2.4M). All generalized figures have 78 vertices (15%).Sinuosities beneath figures are scaled from 1 to 7. Coastline length is shown at upper left. Line weight multiplied by 5.

6.3 Topics for Continuing Research

To make the selection procedure presented above practical and effective in a variety of situations, additional data and control structure modifications will be needed, along with further tests and evaluations. Likewise, there are other additions and refinements to our methods that should be pursued. These are listed below, from the more specific to the more general (and difficult) ones:

1. Assessments of effects of point placement and lookahead. These parameters were not well-studied in our experiments, and more information is needed on the sensitivity of our methods to how their values are set.

2. Experiments using sinuosity-segmented features. Although we created datasets for three different sinuosity regimes, we have yet to generalize each with a variety of methods and parameters to determine the effects of segmentation.

3. Tests with other map data. Some point attributes might only be relevant to certain feature classes, such as roads or rivers. The behavior of our methods in generalizing data other than shorelines needs study, and map data at larger scales should be processed to determine their effective range of resolution.

4, Enabling and using feature-class-specific point attributes. The only point attribute we studied was sinuosity. Several others could also be implemented and assessed, to indicate the presence of forms such as right-angle turns, parallel lines and circular arcs.

5. Reassing measures of generalization performance. Although many common statistical measures of generalization quality seem to be of dubious value, we believe there are a few that can provide insight as well as data that can help in setting parameters. For example, the coefficient of variation of segment length might help to compare different methods and parameters.

6. Automating parameter-setting. Experience with QTM-based line simplification is almost at the point where we can begin to predict what parameter settings produce better results, given the feature class, scale, output resolution and purpose of the map. This must be done if our methods are to be made robust and user-friendly.

7. File processing improvements. Our prototype software now converts geographic coordinates to QTM IDs (and back) every time a file is used. An archival (binary) QTM file format would eliminate a great deal of this overhead, and would also let datasets to be pre-generalized for rapid retrieval across the scales.

8. Data processing benchmarking. Once further optimizations are made, our code should be profiled to identify relative and absolute timings of core functions,

and to generate processing benchmarks using a representative suite of test data.

9. Conditioning output data to remove artifacts. The local nature of our QTM-based generalization methods can produce graphic and topological artifacts such as spikes, self-touching and knots. Algorithms are needed to detect and correct such problems.

10. Improved user interface. Our prototype's mode of operation, being command-line driven and oriented toward batch processing, does not promote rapid exploration of processing strategies or easy assessment of results. A graphic interface and other control mechanisms needs to be put in place, possibly using Java, Perl or TCL/TK in order to keep the prototype platform-independent and accessible over the internet.

11. Implementing spatial indexing strategies. The experiment reported in appendix E shows that QTM IDs can order and index point features fairly efficiently. However, this is still insufficient to handle extended objects. And while minimum bounding triangles (and attractors) can do this, being regular shapes they too have limitations when handling irregularly-shaped features. Additional constructs ought to be designed, implemented using appropriate data structures and database access mechanisms, and their performance assessed for different types and quantities of features.

12. Further work on conflict detection. Our attractor-based model for identifying spatial conflicts needs refinement and testing to calibrate and validate it. Once in place, it should help to provide more specific information about congestion within features as well as conflicts between them. This effort should proceed in parallel with development of spatial indexing strategies.

This is a fairly extensive to-do list, but it is prioritized in a way that will provide initial feedback from minor modifications, which in turn can help to better understand the scope of later, more comprehensive tasks. The list progresses from rounding out performance evaluations through various minor improvements and optimizations to confront significant basic research and software engineering challenges. It is our hope that through circulating this publication some fresh ideas from new collaborators might come to light. This might involve further domain- or application-specific assessments of QTM's utility, leading to new approaches to exploiting its unique or useful properties that can better realize its full potential.

6.4 Looking Back, Looking Ahead

We are just now reaching a point where we can say decisively that QTM is a useful framework for creating multi-resolution spatial databases. Our studies have confirmed many of our expectations that storing digital map data in a hierarchical coordinate system would facilitate its generalization for display across a wide range of scales. From this perspective, our project has been a success.

But another goal for the project was not well met; handling contextual (holistic) generalization problems involving interactions among features. We planned to apply attractors to this problem, and did in fact implement an approach to handling it. This was the attractor-primitive directory described in section 4.2.3. While it works well enough for the purpose of topologically structuring test data, we did not extend it to identify conflicts between features, as additional components were needed for this that needed more design time. We believe the approach can work, but there are many subtleties and special-case scenarios that must be handled before we can say we have a robust conflict detection (to say nothing of a *conflict resolution*) solution.

Despite its apparent complexity and non-intuitive notation, QTM still seems to be a more robust framework than plane coordinates for map generalization and spatial data handling in general, because of its scale- and resolution-sensitive properties, its ability to document positional accuracy and its capacity to perform global and local spatial indexing. While one might think that using QTM would require major commitments from vendors, database designers and users, and new ways of thinking, this is not necessarily true (although a hierarchical view of location can be a useful perspective). By virtue of operating at very low levels in a database, QTM encoding could be made virtually invisible to users, just as many spatial access mechanisms currently are, and just as the details of plane coordinate systems and map projections can be ignored most of the time. The most apparent evidence of QTM's presence might involve changing from rectangular map tiles to triangular ones, which might be noticeable only during interactive map display.

Looking beyond cartographic generalization, QTM can contribute to geospatial technology by improving the documentation of spatial databases and enabling multi-scale analyses of them. More work needs to be done to establish the practical utility of the model for specific applications, and this cannot go on in isolation. For the concepts and techniques described here to mature and diffuse, the need for multi-scale databases must become more widely appreciated, and commitments made to pushing the boudaries of this new area. In terms of GIS data and process modeling, particularly with respect to cartographic requirements, several related developments seem necessary:

1. Source data need to be qualified at the primitive and vertex levels concerning the roles of primitives and specific vertices within them.

2. Sinuosity attributes — or other vertex-specific measures of shape — should be further researched and populated in databases for generalizing map data.

3. Rules need to be formulated regarding different qualities of generalization demanded by different feature classes (but users should be able to modify, augment or override them); work has progressed in this area in recent years.

4. The context of generalization decisions needs to be taken into account; nearby features can and often should influence strategies and tactics. Data models that isolate layers or features can hamper such negotiations; object models may also be insufficient in the absence of structures thar describe proximity relations.

Working on these issues can lead to improving the entire state of the art, not just our own methods, and much valuable progress has already been made. QTM's contribution to map generalization, limited as it currently is, may yet be significant. In itself, QTM does not model much more than positional certainty, resolution and proximity. As map generalization decisions often require additional cartometric criteria and contextual knowledge, QTM alone clearly does not provide a general solution to this set of problems. Its value seems to lie more in describing space and phenomena rather than prescribing specific application solutions, but this contribution alone could be a substantial one. By its multi-resolution nature, by forging connections between spatial resolution and map scale, and by enabling both space- and object-primary generalization strategies to be pursued together, QTM offers a fresh approach to a difficult complex of problems. As two computer scientists with long experience with spatial data handling have written, both space-primary and object-primary approaches have complementary strengths and weaknesses; both are necessary and need to be reconciled better than they have been:

> Although data-driven partitions turn out to be quite complicated to maintain, they are able to cope with skew data reasonably well, while space-driven partitions fail here. For extremely skew data or whenever worst-case guarantees are more important than average behavior, data-driven partitions may be the method of choice. ... Except for such situations, the adaptivity of data-driven partitions will rarely compensate for the burden of their complexity and fragility. More often, combinations of space-driven and data-driven partitions will be appropriate. (Nievergelt and Widmayer 1997: 175-6)

This perspective is mirrored in the roles assigned to QTM facets and attractors, the former being space-driven partitions, the latter data-driven ones, at least as we have described them. It should be obvious from prior chapters that attractor data structures are much more complex than, for example, quadtrees. Indeed a full implementation of attractors was not attempted in this project, principally because of the complexities of designing expressive data structures and robust algorithms to exploit attractor indexing effectively.

Throughout this exposition it has been argued that the benefits of using a hierarchical coordinate system are many, and the costs are acceptable. Some of the costs are in the form of inefficiencies; the overhead associated with encoding and decoding hierarchical coordinates, plus those incurred in re-engineering spatial databases at low levels. There are also costs associated with the additional research and development activities that will be needed to operationalize QTM's spatial indexing capabilities, beyond initial designs and experiments by various researchers.

As chapter 1 explained, it seems probable that as much, if not more, effort will be spent in creating the extra machinery and incurring processing overhead to overcome inherent limitations of plane and geographic coordinates, particularly with regard to handling consequences of variations in positional data quality.

Taken together in all of its facets, QTM is a fundamental technology for abstracting spatial aspects of reality. It is hoped that the logic and potential benefits of using hierarchical coordinates in general, and quaternary triangular mesh in particular, have been made apparent by this publication. Even if they have, perhaps readers will still respond with a shrug: "So what", they might say, "do we care about these matters enough to alter our fundamental spatial concepts and notations for location, ones that have served us well for centuries?" Should they choose not to alter them, they should still try to understand that in recent years, a lot of what we thought we knew about our world and ourselves is being thrown open to doubt, challenged by innovation, and will probably become less and less useful as time goes on. The world is on the threshold of a new millennium, and just about everything in it will continue to change even more rapidly than it recently has.

In our search for solutions, we find ourselves using a globe-spanning hierarchical structure to sort out a slew of tiny geometric details. It is curious that an approach to handling geospatial data that began with a desire to model our planet as a whole has brought us to such a point. The quaternary triangular mesh, in defining a multifaceted polyhedron, with attractors bonding neighboring facets, conjures up images of large-scale global analytic applications such as biodiversity modeling, weather forecasting or tectonic simulations. Yet it is through studying small details contained within particular mesh elements—or small groups of them—that we tend to learn something new about the world. Do we conquer by dividing, or by putting things together? No matter, there is no need to choose...

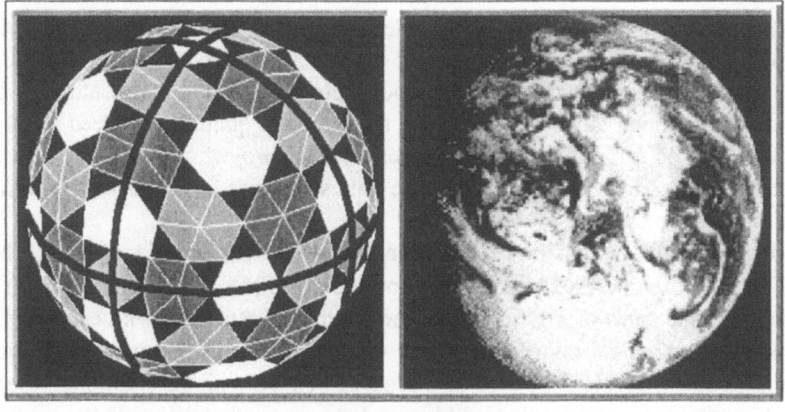

We can't put our planet together, it already is.

Appendix A

Quaternary Triangular Mesh Processing: Fundamental Algorithms

This appendix describes the core algorithms that have been developed by the author to implement QTM data processing, All of them perform certain types of data conversion (to and from different coordinate systems, map generalization operations), or else are utility functions in support of these ends.

Encoding and Decoding Locations: *QTMeval* and related algorithms

Conversion to and from QTM identifiers (QTMIDs) is only possible for data that is or can be expressed as geographic coordinates (latitude and longitude). In addition to such coordinates, some indication of positional certainty for points is also required for correct encoding. This parameter can be global for a dataset, specific to a feature, or may vary from one point to the next, but at least an estimate of it must be supplied. Typically, it is given in units of meters, but other metrics may be used. This parameter's meaning may vary, but that is not of concern to the conversion algorithms described below. For example, the accuracy measure may indicate a 95% probability for the locus of a point, or it may be a circular radius beyond which the point is guaranteed not to lie. The interpretation of the error bound is assumed to be global and data-specific, hence should be recorded as metadata for the encoded dataset. The conversion functions just see it as a distance tolerance.

The approach used by the author for converting geographic locations to QTMIDs relies on an intermediate representation, a map projection called Zenithial OrthoTriangular (ZOT) projection, described below. It transforms an octahedron to a unit square using "Manhattan" distance metric. ZOT coordinates are used as an intermediate notation only; no computations of distances or directions are done using this metric outside of the conversion process. It is employed for two principal reasons: (1) computations in manhattan space require only linear arithmetic; and (2) the QTM hierarchy of triangles is completely regular and uniform in this projection, each facet being a right isosceles triangle that nests completely in a parent triangle of the same shape. This makes conversion relatively fast and efficient.

The core function that converts coordinates is *QTMeval*. It performs conversion in either direction, depending on the arguments it is given: (1) a QTMID is computed for a point in ZOT space, given its x-y coordinates and a linear tolerance that identifies the desired QTM precision; (2) The three ZOT coordinates defining a QTM facet are computed from its QTM ID, limited by a ZOT tolerance. The same code (with one block conditionally executed) does both conversion operations.

Originally, *QTMeval* was implemented as a recursive function that returned a QTMID. An almost identical recursive function produced ZOT coordinates when converting from QTM to ZOT. The two functions were subsequently merged, and now return these values as arguments; the return code is the (integer) maximum QTM level of detail reached before the algorithm halts, or (if negative) an error code. More importantly, *QTMeval* is now iterative rather than recursive. This adds one loop, but no other bookkeeping, to the code. The advantage is that the history of processing facets is preserved across calls, allowing computations to be avoided when successive calls are made to encode or decode points that are relatively close together, as would normally be the case in most applications.

QTMeval's state memory is a global stack that stores 30 detail levels describing the last hierarchy of triangles to be encoded or decoded, implemented as an array of structures passed by the caller. The stack is traversed, and its data used until it is found that the location being encoded or decoded lies outside of one of the QTM sub-facets it defines. From this level on, information in all higher levels of the stack is replaced by data computed for the specific location. The closer that successive points are to one another, the more levels of their stacks are likely to be identical, but this depends on absolute as well as relative positioning wrt the QTM grid. In general, though, this closeness property means that the smaller the area covered by a feature, the more efficient its QTM conversion is likely to be. Only when an input location moves into another QTM Octant must the stack be initialized and recomputed from the beginning; this is a relatively rare event. In describing geographic features, map coordinates weave in and out of QTM quadrants. When they cross into a facet that descends from a different sub-tree, all levels of detail between the final one and that of a common ancestor must be recomputed by *QTMeval*; the stack's memory of those levels is obsolete, and must be refreshed.

Informal trials with a few datasets encoded to between 15 and 20 levels of detail indicate that, on the average, half of all the levels in the stack must be recomputed at each call. This is what one would intuitively expect for medium-scale maps. More densely-digitized maps encoded to higher resolutions might see improved stack utilization levels, but we suspect 50% is a fairly robust statistic. Even though this characteristic doesn't improve *QTMeval*'s order of algorithmic complexity, it does provide a factor-of-two speedup over a recursive implementation. In any case, encoding or decoding QTMIDs for a feature with N vertices to L levels of detail entails a complexity of $O(NL)$.

As *QTMeval* operates in two modes and returns some of its results in a stack, several *wrapper functions* that have cleaner interfaces are used to control its operation. Each one either encodes or decodes data, and handles details such as conversion between geographic and ZOT coordinates, computation of ZOT tolerances, and (in the case of conversion from QTM to coordinates) generation of coordinates to represent a single location for a QTMID (which, as it represents a triangle, not a point, requires guiding assumptions). Implementation details of wrappers are not further discussed here.

The internal logic of *QTMeval* will be described in pseudocode following the code for utility function *getOctant* (which could be a single-line macro if it did not perform bounds-checking). To aid understanding, it will be presented as two functions (*QTMencode* and *QTMdecode*), although they are actually implemented as one. In this presentation, the calling arguments differ in each case for clarity. Two support functions are also listed, *getOctant* and *initStack*. The former identifies which QTM octant a specified ZOT coordinate pair lies in (or returns a diagnostic error code if either coordinate is out of range). The second simply initializes the stack for locations in a specific octant.

```
INTEGER FUNCTION getOctant(FLOAT POINT zxy)

Validates ZOT coordinates;
returns octant (1-8) occupied by zxy or neg error code
Error codes are from -1 to -15, indicating what was out of range.
Inequalities in range tests on x and y determine what octant
points lying on the equator and principal meridians will occupy.

    INTEGER error = 0, oct = 0
    if (zxy.x .GT. 0.)
        if (zxy.x .GT. 1)  error = error - 1
        if (zxy.y .GT. 0.)
            if (zxy.y .GT. 1.)  error = error - 8
            oct = 1
        else
            if (zxy.y .LT. -1.)  error = error - 4
            oct = 2
        end if
    else
        if (zxy.x .LT. -1.)  error = error - 2
        if (zxy.y .LT. 0.)
            if (zxy.y .LT. -1.)  error = error - 4
            oct = 3
        else
            if (zxy.y .GT. 1.)  error = error - 8
            oct = 4
        end if
    end if
    if (error .NE. 0) return error
    In south hemisphere, add 4 to octant number:
    if (|zxy.x| + |zxy.y| .GT. 1.)  oct = oct + 4
    return oct
end getOctant
```

Note: All differences between *QTMencode* and *QTMdecode* are highlighted.

INTEGER FUNCTION *QTMencode* (**CONST FLOAT POINT** zxy,
        CONST FLOAT tol, **INTEGER** lastMatch, **STRUCT** stack[30])

Assume zxy is a valid ZOT coordinate in the range (-1,-1) to (1,1)
Assume tol is a valid ZOT distance in the range $(2^{-29},1.)$
The stack may be unitialized, pristine or contain data

    FLOAT   dx, dy, dz, pn_x, pn_y, xn_x, xn_y, yn_x, yn_y
    INTEGER   oct, QL, pn, xn, yn
    BOOLEAN   SameLoc

    oct = getOctant(zxy) *Compute QTM Octant, oct, occupied by pxy:*
    if (oct .ne. stack[0].QID) initStack (oct, MAXLEVS, stack)
    QL = 0   *Init QTM level*
    SameLoc = *TRUE*   *Init flag for stack's validity*
    dz = 1.0   *Init ZOT x,y edge length (of octant)*
Loop from QTM lvl 0 to the highest that tol will allow:
while (dz .GT. tol)
Get IDs of X- and Y-nodes (xn, yn) from level QL of stack (1,2,3):
    xn = stack[QL].xnode; yn = stack[QL].ynode
    pn = 6 - (xn + yn)   *Compute Polenode ID*
Retrieve ZOT x,y of polenode, pn_x, pn_y from level QL of stack:
    pn_x = stack[0].node_x[pn-1]
    pn_y = stack[0].node_y[pn-1]
    dz = dz / 2   *Reduce closeness criterion by half*
Compute displacement of zxy from polenode:
    dx = |zxy.x - pn_x|
    dy = |zxy.y - pn_y|
Identify closest node to zxy (node 0 represents central facet):
    node = 0
    if (dx + dy .LE. dz) node = pn
    else if (dx .GE. dz) node = xn
    else if (dy .GE. dz) node = yn
    end if
    QL = QL + 1   *Increment QTM level*
Is stack state still consistent at this level?
    if (SameLoc .AND. stack[QL].QID .EQ. node)
        lastMatch = QL
    else   *Need to update rest of stack:*
        SameLoc = FALSE
        xn_x = stack[0].node_x[xn-1]
        xn_y = stack[0].node_y[xn-1]
        yn_x = stack[0].node_x[yn-1]
        yn_y = stack[0].node_y[yn-1]
Copy pn_x, pn_y, xn_x, xn_y, yn_x, yn_y to next level
overriding these values for the two nonselected nodes:
        if (node .NE. pn)
            stack[QL].node_x[pn-1] = pn_x
            stack[QL].node_y[pn-1] = pn_y

```
          else
              stack[QL].node_x[pn-1] = (xn_x + yn_x) / 2
              stack[QL].node_y[pn-1] = (xn_y + yn_y) / 2
          end if
          if (node .NE. yn)
              stack[QL].node_x[yn-1] = yn_x
              stack[QL].node_y[yn-1] = yn_y
          else
              stack[QL].node_x[yn-1] = (xn_x + pn_x) / 2
              stack[QL].node_y[yn-1] = pn_y
          end if
          if (node .NE. xn)
              stack[QL].node_x[xn-1] = xn_x
              stack[QL].node_y[xn-1] = xn_y
          else
              stack[QL].node_x[xn-1] = pn_x
              stack[QL].node_y[xn-1] = (yn_y + pn_y) / 2
          end if
```
Renumber nodes:
```
          if (node .EQ. xn)   yn = 6 - (xn + yn)
          else if (node .EQ. yn)   xn = 6 - (xn + yn)
          else if (node .EQ. pn)
              yn = xn
              xn = 6 - (yn + pn)
          end if
```
Store xnode and ynode ids on stack (pole ID will be derived):
```
          stack[QL}.xnode = xn
          stack[QL}.ynode = yn
          stack[QL].QID = node          Store QID for this level
      end if
  end while
  return QL
```
end *QTMencode*

INTEGER FUNCTION *QTMdecode* (**CONST CHAR** qid[],
  CONST FLOAT tol, **INTEGER** lastMatch, **STRUCT** stack[30])

Assume zxy is a valid ZOT coordinate in the range (-1,-1) to (1,1)
Assume tol is a valid ZOT distance in the range $(2^{-29}, 1.)$
The stack may be unitialized, pristine or contain data

FLOAT   dz, pn_x, pn_y, xn_x, xn_y, yn_x, yn_y
INTEGER   oct, NL, QL, pn, xn, yn
BOOLEAN   SameLoc

NL = *min*(*length*(qid), MAXLEVS)   *QTM levels in qid*
oct = qid[0]
if (oct .ne. stack[0].QID) *initStack* (oct, MAXLEVS, stack)
QL = 0   *Init QTM level*
SameLoc = TRUE   *Init flag for stack's validity*
dz = 1.0   *Init ZOT x,y edge length (of octant)*

Loop from QTM lvl 0 to the highest that tol will allow:
while (dz .GT. tol .AND. QL .LE. NL)
Get IDs of X- and Y-nodes (xn, yn) from lvl QL of stack (1,2,3):
    xn = stack[QL].xnode; yn = stack[QL].ynode
    pn = 6 - (xn + yn)   *Compute Polenode ID*
Retrieve ZOT x,y of polenode from lvl QL of stack:
    pn_x = stack[0].node_x[pn-1]
    pn_y = stack[0].node_y[pn-1]
    dz = dz / 2   *Reduce closeness criterion by half*
    QL = QL + 1   *Increment QTM level*
    node = qid[QL]   *The QTMID specifies the nodes*
Test if stack state is still consistent at this lvl:
    if (SameLoc .AND. stack[QL].QID .EQ. node)
        lastMatch = QL
    else   *Need to update rest of stack:*
        SameLoc = FALSE
        xn_x = stack[0].node_x[xn-1]
        xn_y = stack[0].node_y[xn-1]
        yn_x = stack[0].node_x[yn-1]
        yn_y = stack[0].node_y[yn-1]
Copy pn_x, pn_y, xn_x, xn_y, yn_x, yn_y to next level
overriding these values for the two nonselected nodes:
        if (node .NE. pn)
            stack[QL].node_x[pn-1] = pn_x
            stack[QL].node_y[pn-1] = pn_y
        else
            stack[QL].node_x[pn-1] = (xn_x + yn_x) / 2
            stack[QL].node_y[pn-1] = (xn_y + yn_y) / 2
        end if
        if (node .NE. yn)
            stack[QL].node_x[yn-1] = yn_x
            stack[QL].node_y[yn-1] = yn_y
        else
            stack[QL].node_x[yn-1] = (xn_x + pn_x) / 2
            stack[QL].node_y[yn-1] = pn_y
        end if
        if (node .NE. xn)
            stack[QL].node_x[xn-1] = xn_x
            stack[QL].node_y[xn-1] = xn_y
        else
            stack[QL].node_x[xn-1] = pn_x
            stack[QL].node_y[xn-1] = (yn_y + pn_y) / 2
        end if
        Renumber nodes:
        if (node .EQ. xn) yn = 6 - (xn + yn)
        else if (node .EQ. yn) xn = 6 - (xn + yn)
        else if (node .EQ. pn)
            yn = xn
            xn = 6 - (yn + pn)
        end if

Store xnode and ynode ids on stack(pole ID will be derived):
```
        stack[QL}.xnode = xn
        stack[QL}.ynode = yn
        stack[QL].QID = node              Store QID for this level
    end if
end while
return QL
end QTMdecode
```

PROCEDURE *initStack* (**CONST INTEGER** oct,
                     CONST INTEGER len, **STRUCT** stack[])

Inits stack used by QTMeval for a specified ZOT octant
    Locations for octant vertex positions in ZOT space:
```
    CONST FLOAT zPolenode[8][2] =
        {0.,0.,0.,0.,0.,0.,0.,0.,1.,1.,1.,-1.,-1.,-1, -1.,1.}
    CONST FLOAT zXnode[8][2] =
        {1.,0.,1.,0.,-1.,0.,-1.,0.,0.,1.,0.,-1.,0.,-1.,0.,1.}
    CONST FLOAT zYnode[8][2] =
        {0.,1.,0.,-1.,0.,-1.,0.,1.,1.,0.,1.,0.,-1.,0.,-1.,0.}
    INTEGER i, j, ix, iy, octidx

    octidx = oct-1
    stack[0].QID = oct              First QTM digit is octant num
```
Octant polenodes IDs always = 1, but do not need to be stored
We do need to store the pole coords, however:
```
    stack[0].node_x[0] = zPolenode[octidx][0]
    stack[0].node_y[0] = zPolenode[octidx][1]
    if (oct .LT. 5)                 North hemisphere octants:
        stack[0].xnode = 3                   x-node basis num
        stack[0].ynode = 2                   y-node basis num
        ix = 2; iy = 1                    indices for these
    else                            South hemisphere octants:
        stack[0].xnode = 2                   x-node basis num
        stack[0].ynode = 3                   y-node basis num
    ix = 1; iy = 2                       indices for these
    end if
```
Now that nodes are numbered, install coords for them
```
    stack[0].node_x[ix] = zXnode[octidx][0]
    stack[0].node_y[ix] = zXnode[octidx][1]
    stack[0].node_x[iy] = zYnode[octidx][0]
    stack[0].node_y[iy] = zYnode[octidx][1]
```
Clear rest of stack; QID=9 indicates value not set:
```
    repeat with i = 1 to len-1
        stack[i].QID = 9
        stack[i].xnode = stack[i].ynode = 0
        repeat with j = 0 to 2
            stack[i].node_x[j] = stack[i].node_y[j] = 0.
        end repeat
    end repeat
end initStack
```

Given an Octant ID, *initStack* specifies the three vertices that define this facet. These locations will be used to recursively zero in on a triangular quadrant containing the point being encoded or decoded, a procedure referred to in past work as *trilocation*. At each level of recursion, zero or one of these vertices will remain the same and three or two of them will shift locations.

The six ZOT vertices of the octahedron are stored in tables *zXnode*, *zYnode* and *zPolenode* (summarized below), that are private data to the initializer. These tables encode the three roles a node may play: *Xnodes* (ones that define horizontal edges of facets, *Ynodes* (defining vertical edges), and *Polenodes* (defining the right-angled corner of a facet). Polenodes are so named because they initialy occupy the North and South Poles, and always serve as the local origin for a facet. All higher-level nodes will have the same three basis numbers, but will take on different roles as new facets develop. Their roles change in a regular permutation described below. The IDs and ZOT vertices of the corners of the 8 octants are listed in table A1.

Octant ID	Xnode ID	Xnode X	Xnode Y	Ynode ID	Ynode X	Ynode Y	Pnode ID	Pnode X	Pnode Y
1	3	1.0	0.0	2	0.0	1.0	1	0.0	0.0
2	3	1.0	0.0	2	0.0	-1.0	1	0.0	0.0
3	3	-1.0	0.0	2	0.0	-1.0	1	0.0	0.0
4	3	-1.0	0.0	2	0.0	1.0	1	0.0	0.0
5	2	0.0	1.0	3	1.0	0.0	1	1.0	1.0
6	2	0.0	-1.0	3	1.0	0.0	1	1.0	-1.0
7	2	0.0	-1.0	3	-1.0	0.0	1	-1.0	-1.0
8	2	0.0	1.0	3	-1.0	0.0	1	-1.0	1.0

Table A1: Coordinates of Octahedral Vertices in ZOT Space

Initially, Polenodes are numbered 1. In the Northern Hemisphere, Ynodes are numbered 2, and Xnodes 3 (vice versa in the Southern Hemisphere). These node assignments (also called *Basis Numbers*) cycle in the course of trilocation. Once a node comes into existence and receives a basis number, the number never changes. When new (child) nodes appear (at edge midpoints), their locations are computed as the average of the two existing nodes defining the edge being subdivided. For example, whenever a new QTMID is zero (because a point occupies a central facet) its facet's vertex coordinates are derived as follows:

NewYnode = avg(Xnode, Polenode)
NewXnode = avg(Ynode, Polenode)
NewPolenode = avg(Xnode, Ynode)

where avg() returns the arithmetic mean of the arguments. The basis numbers of new nodes are copied from their parents as follows:

NewYnodeID = XnodeID
NewXnodeID = YnodeID
NewPoleNodeID = PolenodeID

The basis number assigned to a new node is:

1, if occurring between node IDs 2 and 3;
2, if occurring between node IDs 1 and 3;
3, if occurring between node IDs 1 and 2.

That is, the new basis number is 6 - (BN1 + BN2),
where BN1 and BN2 are the basis numbers of nodes forming the bisected edge.

Transitions to QTM IDs other than 0 involve generating two new nodes rather than three. The new nodes will consist either of an Xnode and a Ynode, a Polenode and an Xnode, or a Polenode and a Ynode. The number assigned to the resulting facet is the basis number of the closest existing node to where the point falls (i.e., the node that is *not* new), unless it falls in a central facet (quadrant number zero).

Projective Geometry Algorithms

ZOT (short for Zenithial OrthoTriangular) is a map projection optimized for display and computation of QTM data. This projection maps the 8 facets of an octahedron into a square by unfolding it from the south pole, warping its equilateral facets into right triangles. An algorithm for deriving ZOT from latitude and longitude is listed below, as is the inverse transformation. Projection into ZOT is very efficient, as it involves only linear equations and very few steps. ZOT is not only a computationally simple way to compute QTM addresses, it also can provide a basis for assigning names for the location identifiers we call attractors.

Of course, QTM encoding could be — and has been (Goodchild and Shirin 1989, Otoo and Zhu 1993; Fekete 1990, Lugo and Clarke, 1995) — implemented using other projections as intermediaries; ZOT is not required by the QTM data model. However, this projection has a number of useful properties other than efficiency that can enhance applications built on QTM. In addition, as it is a two-dimensional mapping, computations that use it are easily visualized. Further details concerning this projection may be found in (Dutton 1991a).

In ZOT space all QTM facets (quadrants) are the same shape (isosceles right triangles) and, at a given level, the same size. This is not true for the QTM facets as delineated on the globe, which do vary in shape and size (except at the lowest few levels, the largest facet in each octant is about 1.85 times the size of the smallest one). While the projection is not an equal-area one, it not nearly as nonuniform as many common world projections Alas, it is not conformal either, and is ugly to boot. Its most unique (if not distressing) property is its non-Euclidean distance metric. ZOT distances between points are computed as the sum of the absolute x- and y-distances between them, also known as *Manhattan Metric*.

Prior to determining a QTM address for a geographic location, its coordinates are projected from latitude and longitude to (X,Y) in ZOT space. X and Y coordinates in ZOT run from -1 to 1; the North Pole is centered at (0, 0), and the South Pole occupies the four corners (±1, ±1). The Equator is formed by four interior diagonal octant edges, and contains all points that satisfy the equation: $|x| + |y| = 1$. The Prime Meridian runs vertically through the center, the 90th one horizontally. Northern octants are numbered from 1 to 4, proceeding East from Longitude 0. Southern hemisphere octants are numbered as those directly North of them, plus four. The regularity of the QTM facets in ZOT space and its distance metric are exploited by *QTMeval* to accelerate computations, as explained above. Algorithms for performing ZOT projection (*GeoToZot*) and its inverse (*ZotToGeo*) follow:

PROCEDURE *GeoToZot* (**CONST GEOCOORD** geoPoint, **POINT** zotPoint)

    DOUBLE lon, londeg, dx, dy, dxy, temp

Assume |geoPoint.lat| .LE. 90; Assume |geoPoint.lon| .LE. 180
                                        Get fractional colatitude
    dxy = 1. - (|geoPoint.lat| / 90.)
    lon = geoPoint.lon   *Make longitude positive*
    if (lon .LT. 0.) lon = lon + 360.
    londeg = floor(lon)   *get degree part of longitude*
                                    Normalize lon to (0.-> 90.)
    dx = (londeg MOD 90) + lon - londeg
    dx = dx * (dxy / 90.)   *Derive Manhattan x-distance*
    dy = dxy - dx   *Derive Manhattan y-distance*
    if (geoPoint.lat .LT. 0.)   *Reverse dir. in s. hemi*
        temp = 1. - dx
        dx = 1. - dy
        dy = temp
    end if
                            Swap x&y in octants 2,4,6,8
    if ((floor(lon) / 90) MOD 2 .EQ. 1)
        temp = dx
        dx = dy
        dy = temp
    end if
    Negatize Y value in top half of ZOT space:
    if (lon .GT. 90. .AND. lon .LE. 270.) dy = -dy
    Negatize X value in left half of ZOT space:
    if (lon .GT. 180.) dx = -dx
    zotPoint.x = dx
    zotPoint.y = dy
end GeoToZot

PROCEDURE ZotToGeo(CONST POINT zotPoint, GEOCOORD geoPoint)

```
DOUBLE      dx, dy, dxy, lat, lon
INTEGER     oct;
```

Assume |zotPoint.x| *.LE. 1.0; Assume*|zotPoint.y| *.LE. 1.0*
oct = getOctant(zotPoint)   *octant zot point is in*
dx = |zotPoint.x|
dy = |zotPoint.y|
if (oct .GT. 4)   *Invert x&y in south hemi*
    dx = 1. - dx
    dy = 1. - dy
end if
dxy = dx + dy
if (dxy .EQ. 0.) dxy = 0.00001   *Avoid overflows near poles*
lat = 90. * (1. - dxy)
if (oct .GT. 4)   *Octants 5-8 are in the south*
    lat = -lat
    if (oct MOD 2 .EQ. 1) lon = 90. * dy / dxy
    else lon = 90. * dx /dxy
else   *In the North, do the other thing*
    if (oct MOD 2 .EQ. 1) lon = 90. * dx / dxy
    else lon = 90. * dy /dxy
end if
Offset longitude based on octant number:
lon = lon + ((((oct - 1) MOD 4) * 90) MOD 360)
Negatize west longitudes
if (lon .GT. 180.) lon = lon - 360.
geoPoint.lat = lat
geoPoint.lon = lon
end ZotToGeo

Simplifying Feature Data with QTM: Point Selection Algorithms

As a rule, generalizing spatial data involves a number of operators, such as those formalized by MacMaster and Shea (1992), of which simplification is only one, and line simplification is a subset of that. Analysis of QTM-encoded map data can potentially facilitate other generalization operators, as well as provide a framework for detecting likely conflicts between features at reduced map scales. Chapter 4 describes some of these benefits, which will not be reiterated here. This section describes the technical basis for QTM-based line simplification, as implemented in a function called *weedQids*, which processes one QTM-encoded geometric primitive (arc or polygon) at a time (that is, it does not perform any "contextual" generalization). It then specifies the algorithm, giving a brief walk-through and discussion of it.

This function is called with a list of seven arguments, two of which are data, four of which are calling parameters, and one of which returns a value, as follows:

int   *weedQids* (**int64** qids[ ], **const int64** mels[ ], **const int** npts,
     const int level, **const int** ahead, **const int** hint, **int** nruns)

Arguments are:

qids: Pointer to a list of QTM IDs representing a sequence of vertices along a primitive. One bit of each is used for a deletion flag, which this procedure tests and sets.

mels: Pointer to either a list of (character-encoded) Attractor IDs for each vertex, computed at the resolution of *qids* or less, or a copy of the pointer to *qids*. In either case, there is a 1:1 mapping between *qids* and *mels*. The latter are the spatial partitions used to evaluate the primitive data.

npts: No. of vertices passed in *qids* and *mels*, each the same number of bytes.

level: The QTM level of detail at which to filter *qids*. It must be <= that of the *qids*, and exactly equal to that of *mels* when they are attractors.

ahead: Number of verticies to scan beyond current *mel* (to identify "spikes").

hint: Preferred sinuosity value (encoded in *qids* elements) to use in choosing among vertices, if there is a choice. If 0, ignore such data in QTM IDs.

nruns: Returns the number of separate runs of points (contiguous ones occupying the same element in *mels*) found in processing the primitive.

The function returns the number of points deleted from its input primitive.

No geographic coordinates are provided as data, nor are any references made to them by this algorithm. All decisions are made on the basis of filtering and sampling QTM or attractor identifiers. The list of QTM IDs provided (*qids*) is altered only to set a "hidden" byte in each ID string to mark that vertex for elimination during a subsequent output phase. The list of Attractor IDs (*mels*, which can also be QTM IDs, if desired) is not modified in any way (except trivially, when it is a copy of the pointer to *qids*).

The degree of simplification is primarily governed by the *level* parameter, which defines the size of the QTM attractor or quadrant mesh elements that are analyzed and weeded within. However, the effect of the specified *level* is modulated by the *ahead* parameter (typically in the range of 0 to 10); the larger *ahead* is, the higher the degree of simplification may be (further simplification may be obtained by decreasing *level* or by supplying QTM IDs in *mels* rather than attractor IDs).

Finally, the results of systematically sampling within each *mel* may be overridden by the *hint* parameter, when it is non-zero. To enable this, the value of a reserved byte field in each QTM ID must be set previously. This field contains an integer index describing the degree of local line sinuosity characterizing each vertex (computed from the source data using a local geometric operator).

The operation of the *weedQids* algorithm is summarized next. As mentioned above, it is called once per primitive, and is invoked in either of two modes:

(1) vertex attractors at *level* resolution are provided in *mels*;

(2) vertex QTM IDs at any resolution >= *level* are provided in *mels*;

In mode 2, *mels* is the same memory location as *qids*, and points are filtered by QTM facets rather than by attractor regions. In either case, we refer to the contents of *mels* as mesh elements, which are the primary determinant of the resultant spatial resolution for data generalized by the function.

The function scans mesh element data for a primitive from beginning to end, looking for transitions between *mels* at the scale indicated by *level*. Pointers are used to keep track of which vertices begin and end such sequences ("runs"). Several runs of points may occupy the same mesh element, as the primitive may exit and re-enter it more than once. Such situations can be identified in a limited way by setting calling argument *ahead* to be > 0. Then, that many points beyond the exit point of a run are inspected to see if the run returns to the same mesh element where it began. If so, the run is extended to include the intermediate (<= *ahead*) points that lie outside it.

Once a run of vertices has been identified, one or more of them are selected to represent that set using the integer function *zapPntsInRun*, which determines the number of points to be retained and selects them via systematic sampling. When only one point is retained, it is the median point of the run. If a point is not selected, the "zap" field in its QTM ID in *qids* is set so it can be weeded out later, when the primitive is converted to geographic coordinates for output. Selected points are not modified, although the selection may be overridden, as described below. One additional parameter that strongly affects the degree to simplification is the maximum number of vertices that are permitted to be selected per run. By default one vertex per group of vertices is retained, but more may be kept, up to a user-specified limit. The higher this is, the less simplification will result. This is enforced by *zapPntsInRun*; the user's limit is set via an initialization operation.

If *hint* > 0, it is used as a criterion for fine-tuning point selection. When a point is accepted, its sinuosity value is compared to those of its prior (zapped) neighbors; the one that has a sinuosity value closest to *hint* is instead accepted, if that value is different from the one already chosen. Depending on the value for *hint*, low, high or intermediate values of sinuosity may be favored. Of course, any relevant scaled point attribute information can be substituted for sinuosity.

The pseudocode below presents the algorithm as if the vertex list *qids* and the mesh element list *mels* were encoded as 64-bit integers (which in principle they are). As implemented in the testbed however, these are arrays of characters, and are referenced with pointers rather than array indices. Otherwise (except for syntax) the implementation is the same as the pseudocode. Note that the function simply identifies runs of points within mesh elements, but does no point selection or de-selection. That is handled by a second function, which is described below as well.

```
INTEGER FUNCTION weedQids (int64 qids[],
               const int64 mels[],
               const int npts, const int level,
               const int ahead, const int hint, int nruns)

    int     m1, m2, r1, i, rpts, nzapped

    nzapped = 0
    for (r1=0, m1=2, m2=3; m2 .LT. npts; m2=m2+1)
        if (r1 .EQ. 0)
            if (compareQids(mels[m1], mels[m2], level))
                r1 = m1
                goto NEXTM
            end if
        end if
        if (.NOT. compareQids(mels[m1],mels[m2],level))
            for (i = 1; i .LE. ahead; i = i+1)
                if (compareQids(mels[m2+i],mels[r1],level))
                    m2 = m2+i
                    goto NEXTM
                end if
            end for
            if (r1 .GT. 0)
                rpts = m2-r1
                nzapped = nzapped +
                    zapPntsInRun(qids[r1], rpts, hint)
                r1 = 0
            end if
            nruns = nruns+1
        end if
NEXTM:  m1 = m2
    end for
    if (r1 .GT. 0)
        rpts = m2-r1
        nzapped = nzapped + zapPntsInRun(qids[r1], rpts, hint)
    end if
    return nzapped
end weedQids
```

Several external functions are referred to in the above and other code. They are:

boolean *compareQids*(**int64** id1, **int64** id2, **int** length) — Returns TRUE if *id1* exactly matches *id2* to *length* digits of precision, FALSE otherwise.

int *getHintFrQid*(**int64** qid) — Returns the sinuosity value encoded in *qid* for a vertex, or zero if none was found.

void *zap*(**int64** qid) — Sets a delete flag in QTM ID *qid*.

void *unZap*(**int64** qid) — Clears a delete flag in QTM ID *qid*.

boolean *isZapped*(**int64** qid) — Queries delete flag in QTM ID *qid*, reporting TRUE if it is set, FALSE otherwise.

int *zapPntsInRun*(**int64** qids[ ], **int** rpts, **int** hint) — Returns the number of points in a run of length *rpts* that it eliminated. Refer to pseudocode listing below.

These are simple utilities, except for the point selector function *zapPntsInRun*, described below. The testbed's implementation differs from the pseudocode only in that calculations in the initial *for* loop is replaced by table lookup for efficiency, as the results are entirely determined by run size, not by properties of the data. The inital mechanical selection may then be altered by evaluating vertex sinuosity attributes with respect to the criterion passed in the variable *hint*. This substitution is 1:1, so that the number of selected vertices is not altered.

```
INTEGER FUNCTION zapPntsInRun (int64 qids[],
                               const int rpts,
                               const int hint)
    int     vtx, nzapped, choose, h, bestvtx, besthint
    float   choice, start, spacing

    keep = floor(sqrt((float) rpts))
    nzapped = rpts-keep
    spacing = (float) rpts / (float) keep
    start = 0.5 + (spacing / 2)
    choose = floor(start)
    for (choice=1., vtx=1; vtx .LE. rpts; vtx=vtx+1)
        if (vtx .EQ. choose)
            choose = floor(choice * spacing + start))
            choice = choice+1.
        else zap(qids[vtx])
        end if
    end for
    if (hint .EQ. 0) return (nzapped)
    bestvtx = 0
    besthint = 99999
    for (vtx = 1; vtx .LE. rpts; vtx = vtx+1)
        h = abs(getHintsFrQid(qids[vtx]) - hint)
        if (isZapped(qids[vtx]))
            if (h .LE. besthint)
                bestvtx = vtx
                besthint = h
            end if
        else
            if (besthint .LT. h)
                zap(qids[vtx])
                unZap(qids[bestvtx])
            end if
```

```
            bestvtx = 0
            besthint = 99999
        end if
    end for
    if (besthint .LT. h)
        zap(qids[rpts])
        unZap(qids[bestvtx])
    end if
    return (nzapped)
end zapPntsInRun
```

These two algorithms are simple and fast, their time complexity being linear with the number of points in the primitive plus the number of runs in it times the average size of a run (this amounts to two times the number of points). Maximal simplification results when parameter *ahead* is large (high look-ahead) and *zapPntsInRun* is allowed to retain only one vertex per mesh element (which may be done by limiting the variable *keep* to 1). The degree of reduction also depends on the number of runs per mesh element, which is data-driven (controlled by line complexity). It is also higher when attractors are used as the mesh elements than when QTM facets are. This is because at any given level of detail, attractors on the average have twice the area of facets, and this means that runs of vertices within one tend to be longer; the longer a run is, the greater the degree of data reduction will be.

Besides being data-dependent, the results of the above methods are also sensitive to the settings of three independent parameters in addition to the desired QTM level of detail. Results of some experiments in which these parameters are varied are reported in chapter 4, but these trials certainly do not exhaust the possibilities. For example, if *ahead* is varied from 0 to 3, and the number of points per run is allowed to vary from 1 to 3, there are 12 combinations of these settings, and at least 24 should the effects of specifying *hint* be considered as well.

Finally, despite intricacies of control, these functions are essentially quite simple in concept. While they will never delete the first or last point of a primitive, they do not offer any safeguards against self-intersection — much less intersection of multiple primitives — in their output. To do this would require (rather costly) geometric tests that are not currently possible with QTM-encoded data. If they are deemed necessary, they should be performed on geographic coordinates of primitives generated from lists of selected QTM IDs. As conversion from QTM to latitude and longitude may be done in more than one way (yielding a variety of possible point positions, as section 3.1.3 describes), any line conditioning must by necessity be conducted by geometric post-processing procedures. This situation is no different than that which obtains from applying any of the geometric or statistical line simplification algorithms now in common use; all have the potential for introducing topological problems, especially at high reduction ratios, requiring subsequent cleanup procedures to rectify.

Appendix B

Basic, Computational and Invented Geospatial Terms

The terms and acronyms defined below pertain to geoprocessing in general, not just to the concepts used in this document. Mostly they derive from computational geometry, computer science, geography, cartography and spatial analysis. While many of these terms are relatively common, they may be used in idiosyncratic ways. Others are technical terms — now encountered with increasing frequency — drawn from rather restricted vocabularies within science and mathematics. Still others are unpedigreed appellations of recent provenence, some of which the author pleads guilty of having perpetrated. Amendments, corrections and reactions to these definitions are welcomed.

[Words in *italics* are defined elsewhere in this glossary]

Accuracy (n) The fidelity with which measurements represent the reality they purport to describe.

AID (ack) *Attractor Identifier*; a QTM ID at a certain level of detail that names the attractor containing a given QTM quadrant.

Arc (n) A linear *primitive*, consisting of two or more *points* in a sequence; a *polyline*.

Attractor (n) A *nodal region* that dominates neighboring *cells* in a *tessellation*.

Attribute (n) An identifiable property of or fact about a *location* or an *object*.

Basis Number (n) A *tesseral* digit from 1 up to K, where K is the number of child *cells* of a *basis shape*.

Basis Shape (n) A geometric shape, either a *polygon* or a *polyhedron*, in which a *tessellation* is rooted.

Cell   (n) A *topological primitive*; an areal subdivision of a *tessellation*.

Characteristic Point   (n) A *point* on a *polyline* or along a continuous curve that has perceptual significance in human recognition of a shape.

Coordinate   (n) An ordered n-tuple of numbers that denotes the *position* of a *point* within some *manifold* and metric *space*.

Data   (pl. n) Compiled facts abstracting aspects of a reality of interest; see *trivia*.

Data Model   (n) A conceptual organization of phenomena for the purposes of data abstraction.

Data Quality   (n) The accuracy, consistency, completeness, lineage and currency of a set of measurements; often described as *metadata*.

Data Structure   (n) A specific organization of sets of data elements in a computer's memory.

DLG   (ack) Digital Line Graph: A standard encoding format for vector-topological map data developed by USGS

Dodecahedron   (n) A regular Platonic *polyhedron* consisting of 20 *vertices*, connected by 30 *edges* to form 12 pentagonal *faces*.

Edge   (n) A boundary between *faces* connecting *vertices* of a *graph*; a *one-cell*. ; a *line* defining part of a *polygon*.

Entity   (n) An an identifiable portion or element of physical reality possessing *attributes* and relationships of interest for the purposes of abstracting *data models*.

Face   (n) A *region* in a *graph* bounded by *edges*; a *two-cell*.

Facet   (n) A *cell* in a *tessellation*; e.g. a quadrant of a spherical triangular *quadtree*.

Feature   (n) A set of one or more *primitives* having *locations* and *attributes*, identified as a unique *geographic object*;

(v) to display, advertise or otherwise focus attention on specific entities.

Field

(n) A *tessellation* of a spatial *attribute* describing variation of a single-valued surface.

Forest

(n) A set of trees, specifically of related tree data structures (binary trees, quadtrees, etc). QTM has a forest of eight adjacent quadtrees covering the planet.

Fractal

(n) An n-dimensional geometric object or set that exhibits the properties of *self-similarity* and *fractal dimensionality*.

Fractal Dimension

(n) A measure of geometric complexity of forms; a real number equal to or greater than a form's Euclidean dimensionality.

Fullerene

(n) Carbon molecule having 60 atoms arranged in the form of a truncated *icosahedron*. Named for inventor, architect and geometrician R. Buckminster Fuller.

Genus

(n) The *topological* dimension of a spatial *primitive*.

Geocode

(n) An alphanumeric identifier for a *point, place* or *location* which uniquely identifies what and/or where it is.

Geodata

(n) Contraction of *geographic* data; one or more files of digital information describing geographic phenomena.

Geodesic

(n) The shortest distance between two *points* on a curved surface; (adj) *polyhedral* representation of spheroids.

Geodesy

(n) The science of measuring and *modeling* the size and shape of planets.

Geographic

(adj) Pertaining to the spatial properties of phenomena on the surface of the Earth.

Geography

(n) The science of *modeling* where things are and why.

GIS

(ack) Geographic Information System: A digital facility for storing, retrieving, manipulating and displaying spatial and aspatial facts, usually representing maps and *attributes* of real-world phenomena.

Graph   (n) The network of *faces*, *edges* and *vertices* (*0-*, *1-* and *2-cells*) defined by a set of *points*, *lines* and *polygons*.

Graticule   (n) A regular ruling across a *model* surface; a standard reference *mesh* based on a *grid* oriented in a *manifold*.

Grid   (n) A regularly arrayed constellation of *points*.

Heap   (n) A *data structure* consisting of a contiguous list of variable-length *objects* plus a contiguous list of pointers to addresses where those *objects* are stored.

Hierarchy   (n) A cascading tree or pyramid of *objects* or *locations*, subdividing either regularly or irregularly.

Icosahedron   (n) A regular Platonic *polyhedron* consisting of 12 *vertices*, connected by 30 *edges* to form 20 triangular *faces*.

Identifier   (n) A number or a character string that (uniquely) names an *object* or *location* for general reference; may serve as a search key or *geocode*.

Line   (n) A 1-dimensional geometric *object* that can be embedded in a planar *manifold* and which occupies no area; two or more points connected sequentially.

Linear Quadtree (n) A quadtree *data structure* that records *identifiers* and *attributes* of the leaf nodes of a *quaternary hierarchy*, usually as a pre-ordered list.

Locality   (n) A specific, relatively small *region* of *geographic space*.

Location   (n) A well-defined *region* of *space* that may be thought of as occupying a *point* and explicitly identified by name or *geocode*.

Location Code   (n) See *geocode*.

Manhattan Metric (n) A linear mapping in the Cartesian plane in which distance between points is the sum of x and y distance. Also denoted as IR_1 space.

Manifold   (n) A *model* surface characterized by its *topological* genus.

Map Generalization   (n) The selection and simplified representation of cartographic detail appropriate to the scale and/or purpose of a map.

Map Scale   (n) The ratio of the extent of *features* on a map to the entities on the ground that correspond to them, expressed as 1:n, where n is termed the scale demoninator. Scale need not be constant everywhere in a map.

MBT   (ack) Minimum Bounding Triangle, e.g., the smallest QTM *facet* that contains a given data *object*, such as a *primitive*.

mel   (ack) Mesh ELement; a polygonal *cell* in a *mesh* used for spatial filtering, such as a square, triangle or hexagon.

Mesh   (n) A *graticule* formed by *edges* of a *tessellation*. (a set of *one-cells*)

Metadata   (n) Information compiled to summarize properties or provenance of a dataset; pertinent information omitted from a dataset.

Model   (n) A conceptual image, analog, or physical representation of phenomena; (v) to define the elements of a system and organize them into a cognitive structure.

Nodal Region   (n) An area dominated a *node* in a network or *mesh*.

Node   (n) A *vertex* in a *mesh*; a *point* in a network or a *graph* where *lines* terminate and meet.

Nonnull-cell   (n) A *cell* in a *tessellation* having a nonzero *basis number*.

Null-cell   (n) A *cell* in a *tessellation* having a *basis number* of zero.

Object   (n) An abstraction of a physical *entity* which has an *identifier*, and may have *attributes* as well as a *location* in *space* and time. Objects can have membership in (be instances of) a *class*.

Object-primary   (adj) An approach to geometric computation that explicitly models discrete *phenomena* rather than the continuous *space* they occupy; usually refers to *vector* and *topological* data *models*

Octahedron   (n) A regular Platonic *polyhedron* consisting of 6 *vertices*, connected by 12 *edges* to form 8 triangular *faces* all with equal areas, angles and edge lengths.

Octant   (n) A single facet of an *octahedron*, a spherical right triangle (one eighth of a planet).

One-cell   (n) The *topological* role played by an *edge* in a *graph*.

Place   (n) A *locality* that can be distinquished from all other separate, adjoining and overlapping ones.

Platonic Solids   (n) The set of regular *polyhedra* having identical faces and angles; there are five: tetrahedron, hexahedron, *octahedron, dodecahedron, icosahedron*.

Point   (n) A 0-dimensional *location* in *space*; if connected by *lines* to other *points*, also called a *vertex*.

Polyhedron   (n) A solid bounded by planes and defined by the *vertices* and *edges* where they intersect.

Polygon   (n) A 2-dimensional geometric *object* that can be embedded in a plane *manifold* and which defines a closed *region* or multiple disjoint ones.

Polyline   (n) A 1-dimensional geometric element that can be embedded in a plane *manifold* and which defines a curve using a finite sequence of *vertices*.

Precision   (n) The degree of detail available for a measurement.

Primitive   (n) An simple geometric or *topological object*; a component of a *feature*, such as a *point*, a *polyline* or a simple *polygon*.

Quadtree   (n) A hierarchical *data structure* that keeps track of a *quaternary* division of *space* or *objects*.

Quaternary   (adj) Divided into or pertaining to four parts or things.

Quincunx   (n) Arrangement of a square into five parts or things.

QTM   (ack) Quaternary Triangular Mesh: A hierarchical tri-
                      angular *grid* in which each *cell* may divide into four
                      similar ones.

QTM ID   (abbr) Quaternary Triangular Mesh Identifier: A
                      numeric name for a leaf node in the *QTM region
                      quadtree*. It is a sequence of up to 29 *quaternary* digits
                      preceded by an *octant* ID from 1 through 8.

Range Query   (n) A request to a computer program for the identities
                      of spatial objects that occupy a *region*, usually a
                      rectangle.

Raster   (n) A *grid* of *points*, usually rectangular, within which
                      tessellar data is collected, stored or displayed; (adj) the
                      type of data that scanning devices collect and display.

RDP   (abbr) Ramer-Douglas-Peucker; a global cartographic
                      line simplification algorithm independently discovered
                      ca. 1972 by Urs Ramer and David Douglas.

Region   (n) An identifiable two-dimensional area on a plane or
                      a planet, often represented by *polygons*. A region can
                      be convex or not, and either simply- or multiply-
                      connected.

Segment   (n) Line segment: a straight line connecting two
                      vertices; an *edge* component of a *polyline*, *polygon* or
                      polyhedron.

Self-similarity   (n) A quality of a geometric *object* that causes it to
                      look like itself regardless of the scale of represen-
                      tation; part of the definition of a *fractal*.

Space   (n) A *mesh* of *locations* embedded in a *manifold*.

Space-primary   (adj) An approach to geometric computation that
                      explicitly models *space* rather than *objects* it con-
                      tains; usually refers to *raster* data *models*

Spatial Index   (n) A computed quantity that identifies the location of
                      spatial data items in computer memory, particularly
                      on a mass storage device; a gazetteer.

Surface   (n) A connected *region* or *two-cell*, such as a *polygon*, which may undulate in three dimensions and serve as a container for *objects* and *attributes*.

Tessellar   (adj) Pertaining to properties of and operations on *tessellations*; a form of spatial addressing that substitutes *location* codes for *coordinates*, sometimes organized as a *hierarchy*.

Tessellation   (n) A set of similar or dissimilar shapes that fit together to exhaustively fill *space*, usually in some regular way.

Two-cell   (n) The *topological* role played by a *face* in a *graph*.

Topological   (adj) Properties of *objects* and *space* , such as adjacency and connectedness, that remain invariant under continuous transformations.

Trivia   (pl. n) Compiled facts abstracting aspects of a reality not of interest; see *data*.

Vector   (n) A connection between two *points* in *space*; an *edge* in a *graph*; a direction and distance in *space*; (adj) the type of data that point-addressable devices collect and display.

Vertex   (n; vertices pl.) A mathematical *point* that serves to bound two or more *edges* in a *graph*, locates an inflection point in a *polyline,* or defines a *node* in a *tessellation*.

Voronoi Diagram   (n) A *graph* constructed in N dimensions describing the neighborhoods if a set of *points*, such that all points within any polygonal partition are closer to its seed point than to any other seed point.

Zero-cell   (n) The *topological* role played by a *vertex* in a *graph*.

ZOT   (ack) Zenithial OrthoTriangular; name of a projection mapping an octahedron to a square domain in *Manhattan Metric*.

Appendix C

Generalization Experiments: Statistical Results

Summary of NON-HIERARCHICAL generalization results for: Bainbridge Island WA   Figure C-1.1 (A)

														Scale for	Qlev for
FIG	RUN	LVL	AHD	PPR	TYP	NPTS	%KEPT	TOTKM	SEG-KM	QRES-KM	SCALE(0.4)	seg(mm)	seg/qres	seg(mm)=1	1 mm res
D1.1	0	20				656	100.0	77.4	0.1	0.0095	24,000	4.92	12.39	118,000	17.7
D1.4	7	18	0	1	a	586	89.3	76.9	0.131	0.0381	95,000	1.38	3.44	131,000	17.5
D1.4	7	17	0	1	a	484	73.8	76.3	0.158	0.0763	191,000	0.83	2.07	158,000	17.3
D1.4	7	16	0	1	a	320	48.8	73.7	0.231	0.1526	381,000	0.61	1.51	231,000	16.7
D1.4	7	15	0	1	a	193	29.4	71.3	0.372	0.3052	763,000	0.49	1.22	372,000	16.0
D1.4	7	14	0	1	a	108	16.5	67.0	0.626	0.6104	1,526,000	0.41	1.03	626,000	15.3
D1.4	7	13	0	1	a	57	8.7	61.8	1.103	1.2207	3,052,000	0.36	0.90	1,103,000	14.5
-	7	12	0	1	a	24	3.7	52.7	2.292	2.4414	6,104,000	0.38	0.94	2,292,000	13.4
-	7	11	0	1	a	15	2.3	37.0	2.639	4.8828	12,207,000	0.22	0.54	2,639,000	13.2
-	7	10	0	1	a	3	0.5	12.1	6.060	9.7656	24,414,000	0.25	0.62	6,060,000	12.0
D1.6	8	18	4	1	a	586	89.3	76.9	0.131	0.0381	95,000	1.38	3.44	131,000	17.5
D1.6	8	17	4	1	a	469	71.5	75.9	0.162	0.0763	191,000	0.85	2.12	162,000	17.2
D1.6	8	16	4	1	a	312	47.6	73.5	0.236	0.1526	381,000	0.62	1.55	236,000	16.7
D1.6	8	15	4	1	a	179	27.3	69.7	0.392	0.3052	763,000	0.51	1.28	392,000	16.0
D1.6	8	14	4	1	a	98	14.9	64.4	0.664	0.6104	1,526,000	0.44	1.09	664,000	15.2
D1.6	8	13	4	1	a	53	8.1	60.8	1.169	1.2207	3,052,000	0.38	0.96	1,169,000	14.4
-	8	12	4	1	a	24	3.7	52.7	2.292	2.4414	6,104,000	0.38	0.94	2,292,000	13.4
-	8	11	4	1	a	13	2.0	35.4	2.946	4.8828	12,207,000	0.24	0.60	2,946,000	13.1
-	8	10	4	1	a	3	0.5	12.1	6.060	9.7656	24,414,000	0.25	0.62	6,060,000	12.0
-	5	18	0	4	a	588	89.6	76.9	0.131	0.0381	95,000	1.38	3.43	131,000	17.5
-	5	17	0	4	a	495	75.5	76.4	0.155	0.0763	191,000	0.81	2.03	155,000	17.3
-	5	16	0	4	a	366	55.8	74.4	0.204	0.1526	381,000	0.53	1.34	204,000	16.9
-	5	15	0	4	a	275	41.9	74.0	0.270	0.3052	763,000	0.35	0.88	270,000	16.5
-	5	14	0	4	a	203	30.9	71.3	0.353	0.6104	1,526,000	0.23	0.58	353,000	16.1
-	5	13	0	4	a	141	21.5	68.5	0.489	1.2207	3,052,000	0.16	0.40	489,000	15.6
-	5	12	0	4	a	75	11.4	58.9	0.796	2.4414	6,104,000	0.13	0.33	796,000	14.9
-	5	11	0	4	a	32	4.9	35.8	1.155	4.8828	12,207,000	0.09	0.24	1,155,000	14.4
-	5	10	0	4	a	3	0.5	6.4	3.175	9.7656	24,414,000	0.13	0.33	3,175,000	12.9
-	6	18	4	4	a	588	89.6	76.9	0.131	0.0381	95,000	1.38	3.43	131,000	17.5
-	6	17	4	4	a	485	73.9	76.1	0.157	0.0763	191,000	0.82	2.06	157,000	17.3
-	6	16	4	4	a	360	54.9	73.8	0.206	0.1526	381,000	0.54	1.35	206,000	16.9
-	6	15	4	4	a	264	40.2	73.3	0.279	0.3052	763,000	0.37	0.91	279,000	16.5
-	6	14	4	4	a	193	29.4	70.4	0.367	0.6104	1,526,000	0.24	0.60	367,000	16.1
-	6	13	4	4	a	137	20.9	68.2	0.501	1.2207	3,052,000	0.16	0.41	501,000	15.6
-	6	12	4	4	a	75	11.4	58.9	0.796	2.4414	6,104,000	0.13	0.33	796,000	14.9
-	6	11	4	4	a	28	4.3	35.6	1.317	4.8828	12,207,000	0.11	0.27	1,317,000	14.2
-	6	10	4	4	a	3	0.5	6.4	3.175	9.7656	24,414,000	0.13	0.33	3,175,000	12.9
D1.2	3	18	0	1	q	644	98.2	77.3	0.120	0.0381	95,000	1.27	3.15	120,000	17.7
D1.2	3	17	0	1	q	594	90.5	76.9	0.130	0.0763	191,000	0.68	1.70	130,000	17.6
D1.2	3	16	0	1	q	474	72.3	75.8	0.160	0.1526	381,000	0.42	1.05	160,000	17.3
D1.2	3	15	0	1	q	330	50.3	74.1	0.225	0.3052	763,000	0.30	0.74	225,000	16.8
D1.2	3	14	0	1	q	186	28.4	70.4	0.381	0.6104	1,526,000	0.25	0.62	381,000	16.0
D1.2	3	13	0	1	q	99	15.1	66.8	0.681	1.2207	3,052,000	0.22	0.56	681,000	15.2
-	3	12	0	1	q	48	7.3	60.2	1.281	2.4414	6,104,000	0.21	0.52	1,281,000	14.3
-	3	11	0	1	q	27	4.1	50.0	1.922	4.8828	12,207,000	0.16	0.39	1,922,000	13.7
-	3	10	0	1	q	15	2.3	41.0	2.925	9.7656	24,414,000	0.12	0.30	2,925,000	13.1
-	4	18	4	1	q	644	98.2	77.3	0.120	0.0381	95,000	1.27	3.15	120,000	17.7
-	4	17	4	1	q	591	90.1	76.9	0.130	0.0763	191,000	0.68	1.71	130,000	17.6
-	4	16	4	1	q	470	71.6	75.7	0.161	0.1526	381,000	0.42	1.06	161,000	17.2
-	4	15	4	1	q	318	48.5	73.4	0.232	0.3052	763,000	0.30	0.76	232,000	16.7
-	4	14	4	1	q	174	26.5	69.6	0.402	0.6104	1,526,000	0.26	0.66	402,000	15.9
-	4	13	4	1	q	95	14.5	66.6	0.708	1.2207	3,052,000	0.23	0.58	708,000	15.1
-	4	12	4	1	q	44	6.7	59.2	1.377	2.4414	6,104,000	0.23	0.56	1,377,000	14.1
-	4	11	4	1	q	23	3.5	45.8	2.082	4.8828	12,207,000	0.17	0.43	2,082,000	13.6
-	4	10	4	1	q	13	2.0	37.0	3.079	9.7656	24,414,000	0.13	0.32	3,079,000	13.0
-	1	18	0	4	q	644	98.2	77.3	0.120	0.0381	95,000	1.27	3.15	120,000	17.7
-	1	17	0	4	q	594	90.5	76.9	0.130	0.0763	191,000	0.68	1.70	130,000	17.6
-	1	16	0	4	q	487	74.2	76.1	0.157	0.1526	381,000	0.41	1.03	157,000	17.3
-	1	15	0	4	q	374	57.0	74.8	0.201	0.3052	763,000	0.26	0.66	201,000	16.9
-	1	14	0	4	q	270	41.2	73.7	0.274	0.6104	1,526,000	0.18	0.45	274,000	16.5
-	1	13	0	4	q	199	30.3	71.4	0.361	1.2207	3,052,000	0.12	0.30	361,000	16.1
-	1	12	0	4	q	127	19.4	66.0	0.524	2.4414	6,104,000	0.09	0.21	524,000	15.5
-	1	11	0	4	q	65	9.9	51.7	0.808	4.8828	12,207,000	0.07	0.17	808,000	14.9
-	1	10	0	4	q	29	4.4	38.3	1.369	9.7656	24,414,000	0.06	0.14	1,369,000	14.2
-	2	18	4	4	q	644	98.2	77.3	0.120	0.0381	95,000	1.27	3.15	120,000	17.7
-	2	17	4	4	q	592	90.2	76.9	0.130	0.0763	191,000	0.68	1.71	130,000	17.6
-	2	16	4	4	q	485	73.9	75.9	0.157	0.1526	381,000	0.41	1.03	157,000	17.3
-	2	15	4	4	q	366	55.8	74.2	0.203	0.3052	763,000	0.27	0.67	203,000	16.9
-	2	14	4	4	q	261	39.8	73.4	0.282	0.6104	1,526,000	0.18	0.46	282,000	16.4
-	2	13	4	4	q	196	29.9	71.3	0.366	1.2207	3,052,000	0.12	0.30	366,000	16.1
-	2	12	4	4	q	122	18.6	65.7	0.543	2.4414	6,104,000	0.09	0.22	543,000	15.5
-	2	11	4	4	q	59	9.0	51.3	0.885	4.8828	12,207,000	0.07	0.18	885,000	14.8
-	2	10	4	4	q	27	4.1	38.1	1.466	9.7656	24,414,000	0.06	0.15	1,466,000	14.1

Series Runs SinLvls Min-Sin-Max Smooth Wgts? Method   MinSep mm
1   72   7   1   3   0   Y   NH A/Q   0.4

														Scale for	Qlev for
	Summary of HIERARCHICAL generalization results for:									Bainbridge Island WA				Figure C-1.1 (B)	

Summary of HIERARCHICAL generalization results for: Bainbridge Island WA   *Figure C-1.1 (B)*

Series Runs SinLvls   Min-Sin-Max Smooth Wgts? Method   MinSep mm

1	64	7	1	3	0	Y	HA,HQ			0.4			Scale for	Qlev for
FIG	RUN	LVL	AHD	PPR	TYP	NPTS	%KEPT	TOTKM	SEG-KM	QRES-KM	SCALE(0.4)	seg(mm)	seg/qres	seg(mm)=1	1 mm res
D1.1	0	20				656	100.0	77.4	0.1	0.0095	24,000	4.92	12.39	118,000	17.7
D1.5	7	18	0	1	ha	576	87.8	76.8	0.134	0.0381	95,000	1.41	3.50	134,000	17.5
D1.5	7	17	0	1	ha	439	66.9	75.6	0.173	0.0763	191,000	0.90	2.26	173,000	17.1
D1.5	7	16	0	1	ha	229	34.9	69.6	0.31	0.1526	381,000	0.80	2.00	305,000	16.3
D1.5	7	15	0	1	ha	83	12.7	58.3	0.710	0.3052	763,000	0.93	2.33	710,000	15.1
D1.5	7	14	0	1	ha	23	3.5	38.2	1.737	0.6104	1,526,000	1.14	2.85	1,737,000	13.8
D1.5	7	13	0	1	ha	6	0.9	32.4	6.474	1.2207	3,052,000	2.12	5.30	6,474,000	11.9
-	7	12	0	1	ha	3	0.5	17.6	8.795	2.4414	6,104,000	1.44	3.60	8,795,000	11.5
-	7	11	0	1	ha	2	0.3	1.0	1.000	4.8828	12,207,000	0.08	0.20	1,000,000	14.6
D1.7	8	18	4	1	ha	576	87.8	76.8	0.134	0.0381	95,000	1.41	3.50	134,000	17.5
D1.7	8	17	4	1	ha	429	65.4	74.8	0.175	0.0763	191,000	0.91	2.29	175,000	17.1
D1.7	8	16	4	1	ha	222	33.8	69.3	0.31	0.1526	381,000	0.82	2.05	314,000	16.3
D1.7	8	15	4	1	ha	81	12.3	56.3	0.704	0.3052	763,000	0.92	2.31	704,000	15.1
D1.7	8	14	4	1	ha	23	3.5	40.5	1.839	0.6104	1,526,000	1.21	3.01	1,839,000	13.7
D1.7	8	13	4	1	ha	6	0.9	32.4	6.474	1.2207	3,052,000	2.12	5.30	6,474,000	11.9
-	8	12	4	1	ha	3	0.5	17.6	8.795	2.4414	6,104,000	1.44	3.60	8,795,000	11.5
-	8	11	4	1	ha	2	0.3	1.0	1.000	4.8828	12,207,000	0.08	0.20	1,000,000	14.6
-	5	18	0	4	ha	578	88.1	76.8	0.133	0.0381	95,000	1.40	3.49	133,000	17.5
-	5	17	0	4	ha	447	68.1	75.7	0.170	0.0763	191,000	0.89	2.22	170,000	17.2
-	5	16	0	4	ha	263	40.1	71.3	0.27	0.1526	381,000	0.71	1.78	272,000	16.5
-	5	15	0	4	ha	128	19.5	66.3	0.522	0.3052	763,000	0.68	1.71	522,000	15.5
-	5	14	0	4	ha	53	8.1	51.6	0.992	0.6104	1,526,000	0.65	1.63	992,000	14.6
-	5	13	0	4	ha	13	2.0	34.7	2.893	1.2207	3,052,000	0.95	2.37	2,893,000	13.1
-	5	12	0	4	ha	4	0.6	27.1	9.037	2.4414	6,104,000	1.48	3.70	9,037,000	11.4
-	5	11	0	4	ha	2	0.3	1.0	1.000	4.8828	12,207,000	0.08	0.20	1,000,000	14.6
-	6	18	4	4	ha	578	88.1	76.8	0.133	0.0381	95,000	1.40	3.49	133,000	17.5
-	6	17	4	4	ha	441	67.2	75.5	0.171	0.0763	191,000	0.90	2.25	171,000	17.2
-	6	16	4	4	ha	255	38.9	69.6	0.27	0.1526	381,000	0.72	1.80	274,000	16.5
-	6	15	4	4	ha	123	18.8	63.0	0.516	0.3052	763,000	0.68	1.69	516,000	15.6
-	6	14	4	4	ha	54	8.2	53.4	1.008	0.6104	1,526,000	0.66	1.65	1,008,000	14.6
-	6	13	4	4	ha	14	2.1	36.7	2.821	1.2207	3,052,000	0.92	2.31	2,821,000	13.1
-	6	12	4	4	ha	4	0.6	27.1	9.037	2.4414	6,104,000	1.48	3.70	9,037,000	11.4
-	6	11	4	4	ha	2	0.3	1.0	1.000	4.8828	12,207,000	0.08	0.20	1,000,000	14.6
D1.3	3	18	0	1	hq	644	98.2	77.3	0.120	0.0381	95,000	1.27	3.15	120,000	17.7
D1.3	3	17	0	1	hq	592	90.2	76.9	0.130	0.0763	191,000	0.68	1.71	130,000	17.6
D1.3	3	16	0	1	hq	462	70.4	75.6	0.16	0.1526	381,000	0.43	1.07	164,000	17.2
D1.3	3	15	0	1	hq	294	44.8	71.7	0.245	0.3052	763,000	0.32	0.80	245,000	16.6
D1.3	3	14	0	1	hq	131	20.0	62.5	0.481	0.6104	1,526,000	0.32	0.79	481,000	15.7
D1.3	3	13	0	1	hq	41	6.3	45.8	1.145	1.2207	3,052,000	0.38	0.94	1,145,000	14.4
-	3	12	0	1	hq	14	2.1	35.0	2.688	2.4414	6,104,000	0.44	1.10	2,688,000	13.2
-	3	11	0	1	hq	6	0.9	26.6	5.316	4.8828	12,207,000	0.44	1.09	5,316,000	12.2
-	4	18	4	1	hq	644	98.2	77.3	0.120	0.0381	95,000	1.27	3.15	120,000	17.7
-	4	17	4	1	hq	589	89.8	76.9	0.131	0.0763	191,000	0.68	1.71	131,000	17.5
-	4	16	4	1	hq	458	69.8	75.5	0.17	0.1526	381,000	0.43	1.08	165,000	17.2
-	4	15	4	1	hq	279	42.5	69.7	0.251	0.3052	763,000	0.33	0.82	251,000	16.6
-	4	14	4	1	hq	120	18.3	62.2	0.523	0.6104	1,526,000	0.34	0.86	523,000	15.5
-	4	13	4	1	hq	38	5.8	45.8	1.236	1.2207	3,052,000	0.41	1.01	1,236,000	14.3
-	4	12	4	1	hq	11	1.7	27.8	2.784	2.4414	6,104,000	0.46	1.14	2,784,000	13.1
-	4	11	4	1	hq	4	0.6	13.5	4.493	4.8828	12,207,000	0.37	0.92	4,493,000	12.4
-	1	18	0	4	hq	644	98.2	77.3	0.120	0.0381	95,000	1.27	3.15	120,000	17.7
-	1	17	0	4	hq	592	90.2	76.9	0.130	0.0763	191,000	0.68	1.71	130,000	17.6
-	1	16	0	4	hq	474	72.3	75.9	0.16	0.1526	381,000	0.42	1.05	160,000	17.3
-	1	15	0	4	hq	336	51.2	73.9	0.221	0.3052	763,000	0.29	0.72	221,000	16.8
-	1	14	0	4	hq	198	30.2	69.2	0.351	0.6104	1,526,000	0.23	0.58	351,000	16.1
-	1	13	0	4	hq	94	14.3	62.5	0.672	1.2207	3,052,000	0.22	0.55	672,000	15.2
-	1	12	0	4	hq	36	5.5	47.7	1.362	2.4414	6,104,000	0.22	0.56	1,362,000	14.2
-	1	11	0	4	hq	18	2.7	36.7	2.156	4.8828	12,207,000	0.18	0.44	2,156,000	13.5
-	2	18	4	4	hq	644	98.2	77.3	0.120	0.0381	95,000	1.27	3.15	120,000	17.7
-	2	17	4	4	hq	590	89.9	76.9	0.131	0.0763	191,000	0.68	1.71	131,000	17.5
-	2	16	4	4	hq	472	72.0	75.6	0.16	0.1526	381,000	0.42	1.05	161,000	17.2
-	2	15	4	4	hq	323	49.2	72.8	0.226	0.3052	763,000	0.30	0.74	226,000	16.8
-	2	14	4	4	hq	186	28.4	68.7	0.371	0.6104	1,526,000	0.24	0.61	371,000	16.0
-	2	13	4	4	hq	88	13.4	62.5	0.719	1.2207	3,052,000	0.24	0.59	719,000	15.1
-	2	12	4	4	hq	29	4.4	40.5	1.448	2.4414	6,104,000	0.24	0.59	1,448,000	14.1
-	2	11	4	4	hq	12	1.8	27.2	2.474	4.8828	12,207,000	0.20	0.51	2,474,000	13.3

Figure C-1.2

Bainbridge Island WA

Effects of Specifying Preferred Sinuosity for Non-hierarchical and Hierarchical QTM Simplification

LVL	METHOD	NPTS	PCT	LEN (KM)	SEG (KM)	SEG(MM)	PSV	SIN(1)	SIN(2)	SIN(3)	SIN(4)	SIN(5)	SIN(6)	SIN(7)
14	NHA	131	20	64.7	0.498	0.332	1	33	46	28	13	2	9	0
14	NHA	142	22	55.8	0.396	0.264	7	8	31	27	19	20	24	13
14	HA	68	10	71.6	1.069	0.712	1	24	23	13	4	1	3	0
14	HA	78	12	66.7	0.866	0.577	7	3	9	13	10	14	18	11
15	NHA	210	32	69.4	0.332	0.437	1	35	75	44	35	11	7	3
15	NHA	221	34	62.3	0.283	0.373	7	22	56	38	40	22	31	12
15	HA	122	19	73.1	0.604	0.795	1	25	43	30	17	4	2	1
15	HA	128	20	69.0	0.543	0.715	7	13	32	19	19	13	23	9
16	NHA	332	51	73.0	0.221	0.580	1	43	117	77	55	20	16	4
16	NHA	335	51	74.3	0.222	0.585	7	39	108	68	51	30	26	13
16	HA	256	39	70.1	0.275	0.723	1	41	101	54	36	12	10	2
16	HA	267	41	72.5	0.273	0.717	7	35	92	49	37	21	22	11

PCT: % pts retained LEN: perimeter SEG: Avg. seg. len. (ground & map) SIN: Sinuosity class counts
L: QTM Level H: Hierarchical NH: Non-hierarchical A: Attr. Method PSV: Preferred Sinuosity Value

Distribution of Sinuosity Values after generalization (SIN(1) - SIN(7) above)
Histograms are arranged as maps in Figure D-1.11

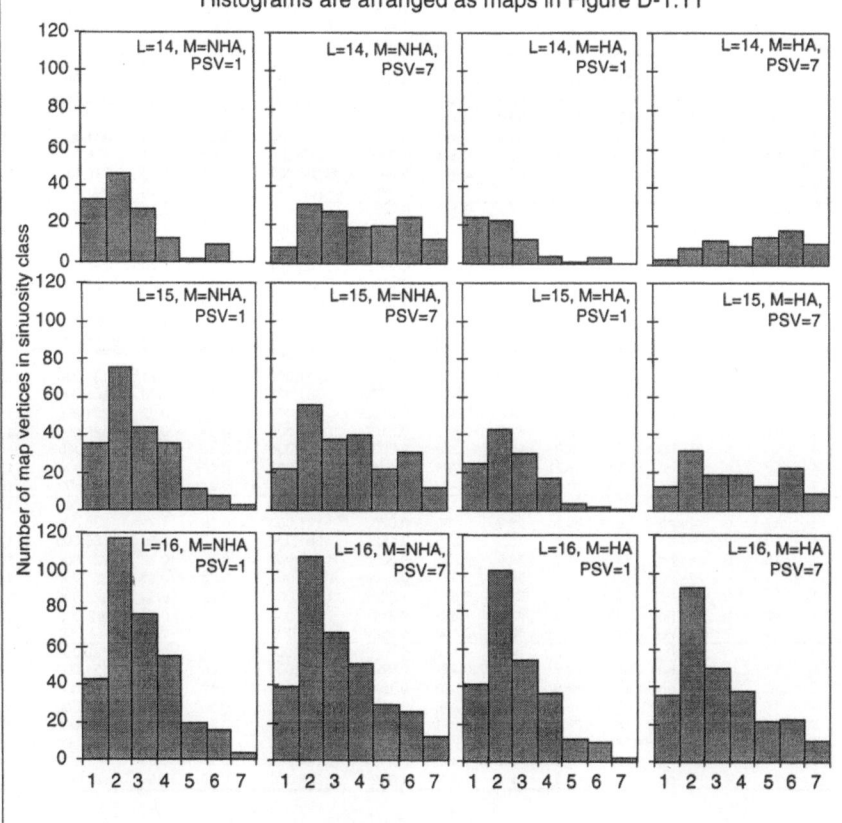

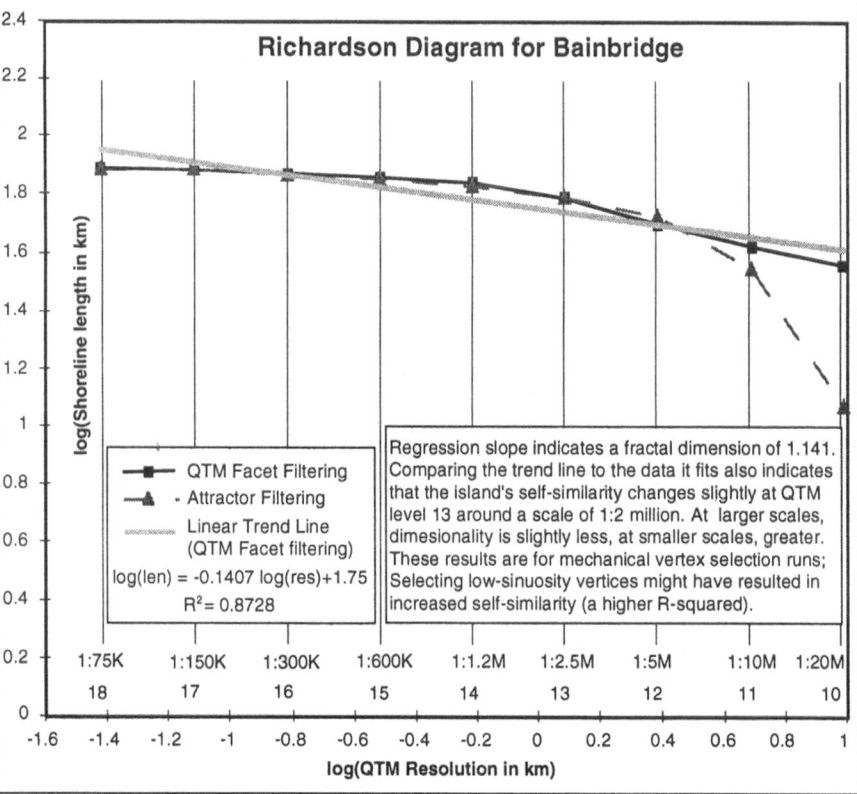

Figure C-1.3

Bainbridge Island WA: Length Statistics for Coastline

Bainbridge Island WA: Length Statistics for QTM Facet Filtering

QTM LEVEL	18	17	16	15	14	13	12	11	10
VERTICES	577	541	451	320	184	97	46	25	13
LENGTH	77.6	77.1	75.2	72.8	69	62	49.9	42	36.2
SEG_LEN KM	0.135	0.143	0.167	0.228	0.377	0.646	1.109	1.750	3.017
QTM_RES KM	0.038	0.076	0.153	0.305	0.610	1.221	2.441	4.883	9.766
LOG(LENGTH)	1.890	1.887	1.876	1.862	1.839	1.792	1.698	1.623	1.559
LOG(SEG_LEN)	-0.871	-0.845	-0.777	-0.642	-0.424	-0.190	0.045	0.243	0.480
LOG(QTM_RES)	-1.419	-1.118	-0.816	-0.515	-0.214	0.087	0.388	0.689	0.990

Bainbridge Island WA: Length Statistics for Attractor Filtering

QTM LEVEL	18	17	16	15	14	13	12	11	10
VERTICES	576	457	313	189	108	57	24	15	3
LENGTH	77.5	76.5	73.7	71.8	67.5	61.9	53.6	35	11.8
SEG_LEN KM	0.135	0.168	0.236	0.382	0.631	1.105	2.330	2.500	5.900
QTM_RES KM	0.038	0.076	0.153	0.305	0.610	1.221	2.441	4.883	9.766
LOG(LENGTH)	1.889	1.884	1.867	1.856	1.829	1.792	1.729	1.544	1.072
LOG(SEG_LEN)	-0.870	-0.775	-0.627	-0.418	-0.200	0.044	0.367	0.398	0.771
LOG(QTM_RES)	-1.419	-1.118	-0.816	-0.515	-0.214	0.087	0.388	0.689	0.990

Richardson Diagram for Bainbridge

Legend:
- QTM Facet Filtering
- Attractor Filtering
- Linear Trend Line (QTM Facet filtering)

$\log(len) = -0.1407 \log(res) + 1.75$
$R^2 = 0.8728$

Regression slope indicates a fractal dimension of 1.141. Comparing the trend line to the data it fits also indicates that the island's self-similarity changes slightly at QTM level 13 around a scale of 1:2 million. At larger scales, dimensionality is slightly less, at smaller scales, greater. These results are for mechanical vertex selection runs; Selecting low-sinuosity vertices might have resulted in increased self-similarity (a higher R-squared).

y-axis: log(Shoreline length in km)

x-axis (scale / level):
1:75K (18), 1:150K (17), 1:300K (16), 1:600K (15), 1:1.2M (14), 1:2.5M (13), 1:5M (12), 1:10M (11), 1:20M (10)

x-axis: log(QTM Resolution in km) (-1.6 to 1)

FIG	RUN	LVL	AHD	PPR	TYP	NPTS	%KEPT	TOTKM	SEG-KM	QRES-KM	SCALE(0.4)	seg(mm)	seg/qres	Scale for seg(mm)=1	Qlev for 1 mm res
														Summary of NON-HIERARCHICAL generalization results for: Nantucket MA Islands — Figure C-2.1 (A)	
														Series Runs SinLvls Min-Sin-Max Smooth Wgts? Method — MinSep mm	
														1 72 7 1 3 0 Y NH A/Q — 0.4	
D2.1	0	20				1105	100.0	152.0	0.138	0.0095	24,000	5.74	14.44	138,000	17.5
D2.4	7	18	0	1	a	1000	90.5	151.2	0.151	0.0381	95,000	1.59	3.97	151,000	17.3
D2.4	7	17	0	1	a	788	71.3	149.5	0.190	0.0763	191,000	0.99	2.49	190,000	17.0
D2.4	7	16	0	1	a	561	50.8	146.2	0.261	0.1526	381,000	0.68	1.71	261,000	16.5
D2.4	7	15	0	1	a	385	34.8	142.5	0.371	0.3052	763,000	0.49	1.22	371,000	16.0
D2.4	7	14	0	1	a	176	15.9	128.0	0.731	0.6104	1,526,000	0.48	1.20	731,000	15.1
D2.4	7	13	0	1	a	103	9.3	116.0	1.137	1.2207	3,052,000	0.37	0.93	1,137,000	14.4
.	7	12	0	1	a	57	5.2	103.5	1.848	2.4414	6,104,000	0.30	0.76	1,848,000	13.7
.	7	11	0	1	a	37	3.3	100.8	2.799	4.8828	12,207,000	0.23	0.57	2,799,000	13.1
.	7	10	0	1	a	25	2.3	58.0	2.417	9.7656	24,414,000	0.10	0.25	2,417,000	13.3
D2.6	8	18	4	1	a	995	90.0	151.1	0.152	0.0381	95,000	1.60	3.99	152,000	17.3
D2.6	8	17	4	1	a	770	69.7	148.7	0.193	0.0763	191,000	1.01	2.53	193,000	17.0
D2.6	8	16	4	1	a	544	49.2	145.4	0.268	0.1526	381,000	0.70	1.76	268,000	16.5
D2.6	8	15	4	1	a	359	32.5	140.1	0.391	0.3052	763,000	0.51	1.28	391,000	16.0
D2.6	8	14	4	1	a	160	14.5	124.3	0.782	0.6104	1,526,000	0.51	1.28	782,000	15.0
D2.6	8	13	4	1	a	91	8.2	114.2	1.268	1.2207	3,052,000	0.42	1.04	1,268,000	14.3
.	8	12	4	1	a	51	4.6	101.9	2.038	2.4414	6,104,000	0.33	0.83	2,038,000	13.6
.	8	11	4	1	a	33	3.0	100.5	3.141	4.8828	12,207,000	0.26	0.64	3,141,000	13.0
.	8	10	4	1	a	23	2.1	57.4	2.610	9.7656	24,414,000	0.11	0.27	2,610,000	13.2
.	5	18	0	4	a	1000	90.5	151.2	0.151	0.0381	95,000	1.59	3.97	151,000	17.3
.	5	17	0	4	a	805	72.9	149.8	0.186	0.0763	191,000	0.98	2.44	186,000	17.0
.	5	16	0	4	a	628	56.8	147.4	0.235	0.1526	381,000	0.62	1.54	235,000	16.7
.	5	15	0	4	a	496	44.9	145.7	0.294	0.3052	763,000	0.39	0.96	294,000	16.4
.	5	14	0	4	a	321	29.0	140.7	0.440	0.6104	1,526,000	0.29	0.72	440,000	15.8
.	5	13	0	4	a	220	19.9	127.0	0.580	1.2207	3,052,000	0.19	0.48	580,000	15.4
.	5	12	0	4	a	125	11.3	106.7	0.860	2.4414	6,104,000	0.14	0.35	860,000	14.8
.	5	11	0	4	a	93	8.4	110.7	1.204	4.8828	12,207,000	0.10	0.25	1,204,000	14.3
.	5	10	0	4	a	54	4.9	59.5	1.123	9.7656	24,414,000	0.05	0.11	1,123,000	14.4
.	6	18	4	4	a	996	90.1	151.2	0.152	0.0381	95,000	1.60	3.98	152,000	17.3
.	6	17	4	4	a	794	71.9	149.0	0.188	0.0763	191,000	0.98	2.46	188,000	17.0
.	6	16	4	4	a	618	55.9	147.3	0.239	0.1526	381,000	0.63	1.56	239,000	16.7
.	6	15	4	4	a	475	43.0	143.9	0.304	0.3052	763,000	0.40	0.99	304,000	16.3
.	6	14	4	4	a	298	27.0	137.4	0.462	0.6104	1,526,000	0.30	0.76	462,000	15.7
.	6	13	4	4	a	207	18.7	125.5	0.609	1.2207	3,052,000	0.20	0.50	609,000	15.3
.	6	12	4	4	a	119	10.8	106.1	0.899	2.4414	6,104,000	0.15	0.37	899,000	14.8
.	6	11	4	4	a	86	7.8	110.0	1.294	4.8828	12,207,000	0.11	0.27	1,294,000	14.3
.	6	10	4	4	a	50	4.5	58.9	1.202	9.7656	24,414,000	0.05	0.12	1,202,000	14.3
D2.2	3	18	0	1	q	1089	98.6	151.9	0.140	0.0381	95,000	1.47	3.66	140,000	17.4
D2.2	3	17	0	1	q	999	90.4	151.1	0.151	0.0763	191,000	0.79	1.98	151,000	17.3
D2.2	3	16	0	1	q	822	74.4	149.4	0.182	0.1526	381,000	0.48	1.19	182,000	17.1
D2.2	3	15	0	1	q	580	52.5	146.3	0.253	0.3052	763,000	0.33	0.83	253,000	16.6
D2.2	3	14	0	1	q	340	30.8	138.7	0.409	0.6104	1,526,000	0.27	0.67	409,000	15.9
D2.2	3	13	0	1	q	205	18.6	130.4	0.639	1.2207	3,052,000	0.21	0.52	639,000	15.3
.	3	12	0	1	q	119	10.8	119.7	1.015	2.4414	6,104,000	0.17	0.42	1,015,000	14.6
.	3	11	0	1	q	74	6.7	111.9	1.533	4.8828	12,207,000	0.13	0.31	1,533,000	14.0
.	3	10	0	1	q	45	4.1	93.7	2.130	9.7656	24,414,000	0.09	0.22	2,130,000	13.5
.	4	18	4	1	q	1089	98.6	151.9	0.140	0.0381	95,000	1.47	3.66	140,000	17.4
.	4	17	4	1	q	999	90.4	151.1	0.151	0.0763	191,000	0.79	1.98	151,000	17.3
.	4	16	4	1	q	808	73.1	148.8	0.184	0.1526	381,000	0.48	1.21	184,000	17.1
.	4	15	4	1	q	547	49.5	143.9	0.264	0.3052	763,000	0.35	0.86	264,000	16.5
.	4	14	4	1	q	316	28.6	136.3	0.433	0.6104	1,526,000	0.28	0.71	433,000	15.8
.	4	13	4	1	q	185	16.7	127.8	0.695	1.2207	3,052,000	0.23	0.57	695,000	15.1
.	4	12	4	1	q	103	9.3	117.7	1.153	2.4414	6,104,000	0.19	0.47	1,153,000	14.4
.	4	11	4	1	q	62	5.6	110.4	1.809	4.8828	12,207,000	0.15	0.37	1,809,000	13.8
.	4	10	4	1	q	39	3.5	92.2	2.427	9.7656	24,414,000	0.10	0.25	2,427,000	13.3
.	1	18	0	4	q	1089	98.6	151.9	0.140	0.0381	95,000	1.47	3.66	140,000	17.4
.	1	17	0	4	q	999	90.4	151.1	0.151	0.0763	191,000	0.79	1.98	151,000	17.3
.	1	16	0	4	q	835	75.6	149.6	0.179	0.1526	381,000	0.47	1.18	179,000	17.1
.	1	15	0	4	q	649	58.7	147.9	0.228	0.3052	763,000	0.30	0.75	228,000	16.7
.	1	14	0	4	q	471	42.6	144.6	0.308	0.6104	1,526,000	0.20	0.50	308,000	16.3
.	1	13	0	4	q	356	32.2	141.1	0.397	1.2207	3,052,000	0.13	0.33	397,000	15.9
.	1	12	0	4	q	263	23.8	137.0	0.523	2.4414	6,104,000	0.09	0.21	523,000	15.5
.	1	11	0	4	q	192	17.4	128.4	0.672	4.8828	12,207,000	0.06	0.14	672,000	15.2
.	1	10	0	4	q	101	9.1	99.8	0.998	9.7656	24,414,000	0.04	0.10	998,000	14.6
.	2	18	4	4	q	1089	98.6	151.7	0.139	0.0381	95,000	1.47	3.66	139,000	17.5
.	2	17	4	4	q	999	90.4	151.1	0.151	0.0763	191,000	0.79	1.98	151,000	17.3
.	2	16	4	4	q	826	74.8	149.4	0.181	0.1526	381,000	0.48	1.19	181,000	17.1
.	2	15	4	4	q	628	56.8	148.6	0.234	0.3052	763,000	0.31	0.77	234,000	16.7
.	2	14	4	4	q	453	41.0	143.6	0.318	0.6104	1,526,000	0.21	0.52	318,000	16.3
.	2	13	4	4	q	336	30.4	140.2	0.418	1.2207	3,052,000	0.14	0.34	418,000	15.9
.	2	12	4	4	q	244	22.1	135.4	0.557	2.4414	6,104,000	0.09	0.23	557,000	15.5
.	2	11	4	4	q	177	16.0	126.6	0.719	4.8828	12,207,000	0.06	0.15	719,000	15.1
.	2	10	4	4	q	92	8.3	98.7	1.084	9.7656	24,414,000	0.04	0.11	1,084,000	14.5

Summary of HIERARCHICAL generalization results for:								Nantucket MA Islands						*Figure C-2.1 (B)*	
SeriesRuns SinLvls			Min-Sin-Max	Smooth Wgts?	Method			MinSep mm							
1	72	7	1	3	0	Y	HA/HQ			0.4				Scale for	Qlev for
FIG	RUN	LVL	AHD	PPR	TYP	NPTS	%KEPT	TOTKM	SEG-KM	QRES-KM	SCALE(0.4)	seg(mm)	seg/qres	seg(mm)=1	1 mm res
D2.1	0	20				1105	100.0	152.0	0.138	0.0095	24,000	5.74	14.44	138,000	17.5
D2.5	7	18	0	1	ha	980	88.7	150.8	0.154	0.0381	95,000	1.62	4.04	154,000	17.3
D2.5	7	17	0	1	ha	721	65.2	147.3	0.205	0.0763	191,000	1.07	2.68	205,000	16.9
D2.5	7	16	0	1	ha	412	37.3	137.1	0.334	0.1526	381,000	0.88	2.19	334,000	16.2
D2.5	7	15	0	1	ha	207	18.7	120.6	0.585	0.3052	763,000	0.77	1.92	585,000	15.4
D2.5	7	14	0	1	ha	63	5.7	92.2	1.487	0.6104	1,526,000	0.97	2.44	1,487,000	14.0
D2.5	7	13	0	1	ha	16	1.4	55.0	3.669	1.2207	3,052,000	1.20	3.01	3,669,000	12.7
.	7	12	0	1	ha	7	0.6	5.9	0.975	2.4414	6,104,000	0.16	0.40	975,000	14.6
.	7	11	0	1	ha	7	0.6	19.8	3.293	4.8828	12,207,000	0.27	0.67	3,293,000	12.9
D2.7	8	18	4	1	ha	975	88.2	150.7	0.155	0.0381	95,000	1.63	4.06	155,000	17.3
D2.7	8	17	4	1	ha	971	87.9	146.4	0.151	0.0763	191,000	0.79	1.98	151,000	17.3
D2.7	8	16	4	1	ha	395	35.7	134.5	0.341	0.1526	381,000	0.90	2.24	341,000	16.2
D2.7	8	15	4	1	ha	198	17.9	117.4	0.596	0.3052	763,000	0.78	1.95	596,000	15.4
D2.7	8	14	4	1	ha	60	5.4	86.7	1.469	0.6104	1,526,000	0.96	2.41	1,469,000	14.1
D2.7	8	13	4	1	ha	15	1.4	55.0	3.931	1.2207	3,052,000	1.29	3.22	3,931,000	12.6
.	8	12	4	1	ha	7	0.6	5.9	0.975	2.4414	6,104,000	0.16	0.40	975,000	14.6
.	8	11	4	1	ha	7	0.6	19.8	3.293	4.8828	12,207,000	0.27	0.67	3,293,000	12.9
.	5	18	0	4	ha	980	88.7	150.8	0.154	0.0381	95,000	1.62	4.04	154,000	17.3
.	5	17	0	4	ha	732	66.2	147.1	0.201	0.0763	191,000	1.05	2.64	201,000	16.9
.	5	16	0	4	ha	447	40.5	138.1	0.310	0.1526	381,000	0.81	2.03	310,000	16.3
.	5	15	0	4	ha	231	20.9	123.6	0.538	0.3052	763,000	0.70	1.76	538,000	15.5
.	5	14	0	4	ha	76	6.9	97.6	1.301	0.6104	1,526,000	0.85	2.13	1,301,000	14.2
.	5	13	0	4	ha	25	2.3	56.9	2.370	1.2207	3,052,000	0.78	1.94	2,370,000	13.4
.	5	12	0	4	ha	11	1.0	40.8	4.080	2.4414	6,104,000	0.67	1.67	4,080,000	12.6
.	5	11	0	4	ha	7	0.6	19.8	3.293	4.8828	12,207,000	0.27	0.67	3,293,000	12.9
.	6	18	4	4	ha	976	88.3	150.7	0.155	0.0381	95,000	1.63	4.05	155,000	17.3
.	6	17	4	4	ha	722	65.3	146.4	0.203	0.0763	191,000	1.06	2.66	203,000	16.9
.	6	16	4	4	ha	433	39.2	137.0	0.317	0.1526	381,000	0.83	2.08	317,000	16.3
.	6	15	4	4	ha	216	19.5	120.9	0.562	0.3052	763,000	0.74	1.84	562,000	15.4
.	6	14	4	4	ha	73	6.6	92.2	1.281	0.6104	1,526,000	0.84	2.10	1,281,000	14.3
.	6	13	4	4	ha	24	2.2	56.8	2.470	1.2207	3,052,000	0.81	2.02	2,470,000	13.3
.	6	12	4	4	ha	11	1.0	40.8	4.080	2.4414	6,104,000	0.67	1.67	4,080,000	12.6
.	6	11	4	4	ha	7	0.6	19.8	3.293	4.8828	12,207,000	0.27	0.67	3,293,000	12.9
D2.3	3	18	0	1	hq	1089	98.6	151.9	0.140	0.0381	95,000	1.47	3.66	140,000	17.4
D2.3	3	17	0	1	hq	998	90.3	151.0	0.151	0.0763	191,000	0.79	1.99	151,000	17.3
D2.3	3	16	0	1	hq	784	71.0	147.5	0.188	0.1526	381,000	0.49	1.23	188,000	17.0
D2.3	3	15	0	1	hq	521	47.1	140.3	0.270	0.3052	763,000	0.35	0.88	270,000	16.5
D2.3	3	14	0	1	hq	251	22.7	119.8	0.479	0.6104	1,526,000	0.31	0.79	479,000	15.7
D2.3	3	13	0	1	hq	110	10.0	104.8	0.962	1.2207	3,052,000	0.32	0.79	962,000	14.7
.	3	12	0	1	hq	44	4.0	86.5	2.011	2.4414	6,104,000	0.33	0.82	2,011,000	13.6
.	3	11	0	1	hq	18	1.6	46.4	2.730	4.8828	12,207,000	0.22	0.56	2,730,000	13.2
.	4	18	4	1	hq	1089	98.6	143.4	0.132	0.0381	95,000	1.39	3.45	132,000	17.5
.	4	17	4	1	hq	998	90.3	151.0	0.151	0.0763	191,000	0.79	1.99	151,000	17.3
.	4	16	4	1	hq	784	71.0	147.5	0.188	0.1526	381,000	0.49	1.23	188,000	17.0
.	4	15	4	1	hq	486	44.0	137.4	0.283	0.3052	763,000	0.37	0.93	283,000	16.4
.	4	14	4	1	hq	230	20.8	115.5	0.504	0.6104	1,526,000	0.33	0.83	504,000	15.6
.	4	13	4	1	hq	93	8.4	99.1	1.077	1.2207	3,052,000	0.35	0.88	1,077,000	14.5
.	4	12	4	1	hq	30	2.7	56.3	1.942	2.4414	6,104,000	0.32	0.80	1,942,000	13.7
.	4	11	4	1	hq	10	0.9	22.0	2.449	4.8828	12,207,000	0.20	0.50	2,449,000	13.3
.	1	18	0	4	hq	1018	92.1	151.9	0.149	0.0381	95,000	1.57	3.92	149,000	17.4
.	1	17	0	4	hq	998	90.3	151.0	0.151	0.0763	191,000	0.79	1.99	151,000	17.3
.	1	16	0	4	hq	806	72.9	148.0	0.184	0.1526	381,000	0.48	1.20	184,000	17.1
.	1	15	0	4	hq	571	51.7	142.0	0.249	0.3052	763,000	0.33	0.82	249,000	16.6
.	1	14	0	4	hq	327	29.6	132.8	0.407	0.6104	1,526,000	0.27	0.67	407,000	15.9
.	1	13	0	4	hq	196	17.7	121.2	0.621	1.2207	3,052,000	0.20	0.51	621,000	15.3
.	1	12	0	4	hq	108	9.8	110.1	1.029	2.4414	6,104,000	0.17	0.42	1,029,000	14.6
.	1	11	0	4	hq	51	4.6	96.1	1.922	4.8828	12,207,000	0.16	0.39	1,922,000	13.7
.	2	18	4	4	hq	1089	98.6	151.9	0.140	0.0381	95,000	1.47	3.66	140,000	17.4
.	2	17	4	4	hq	998	90.3	151.0	0.151	0.0763	191,000	0.79	1.99	151,000	17.3
.	2	16	4	4	hq	795	71.9	147.4	0.186	0.1526	381,000	0.49	1.22	186,000	17.0
.	2	15	4	4	hq	544	49.2	140.4	0.259	0.3052	763,000	0.34	0.85	259,000	16.6
.	2	14	4	4	hq	308	27.9	129.9	0.423	0.6104	1,526,000	0.28	0.69	423,000	15.9
.	2	13	4	4	hq	173	15.7	113.7	0.661	1.2207	3,052,000	0.22	0.54	661,000	15.2
.	2	12	4	4	hq	86	7.8	107.7	1.268	2.4414	6,104,000	0.21	0.52	1,268,000	14.3
.	2	11	4	4	hq	38	3.4	89.0	2.406	4.8828	12,207,000	0.20	0.49	2,406,000	13.3

Figure C-2.2

Nantucket Island MA

Effects of Specifying Preferred Sinuosity for Non-hierarchical and Hierarchical QTM Line Simplification

LVL	METHOD	NPTS	PCT	LEN (KM)	SEG (KM)	SEG(MM)	PSV	SIN(1)	SIN(2)	SIN(3)	SIN(4)	SIN(5)	SIN(6)	SIN(7)
14	NHA	202	18	125	0.622	0.415	1	64	56	15	10	8	5	3
14	NHA	231	21	137.1	0.596	0.397	7	46	36	16	17	22	21	27
14	HA	111	10	109	0.991	0.661	1	43	36	4	1	3	2	0
14	HA	138	13	125.8	0.918	0.612	7	24	23	9	8	16	12	20
15	NHA	408	37	138.3	0.340	0.447	1	104	85	54	42	24	11	10
15	NHA	434	39	145.1	0.335	0.441	7	98	69	39	32	38	37	36
15	HA	259	23	125.2	0.485	0.639	1	80	58	34	24	12	4	0
15	HA	291	26	138.8	0.479	0.630	7	72	46	22	20	23	25	29
16	NHA	571	52	143.8	0.252	0.664	1	134	106	60	59	55	31	10
16	NHA	593	54	147.4	0.249	0.655	7	128	92	64	52	54	52	29
16	HA	448	41	137.1	0.307	0.807	1	125	92	52	40	40	14	1
16	HA	487	44	144.1	0.297	0.780	7	118	81	54	39	37	36	27

PCT: % pts retained   LEN: perimeter   SEG: Avg. segment length (ground & map)   SIN: Sinuosity class counts
L: QTM Level   H: Hierarchical   NH: Non-hierarchical   A: Attractor Method   PSV: Preferred Sinuosity Value

Distribution of Sinuosity Values after generalization (SIN(1) - SIN(7) above)
Histograms are arranged as maps in Figure D-1.8

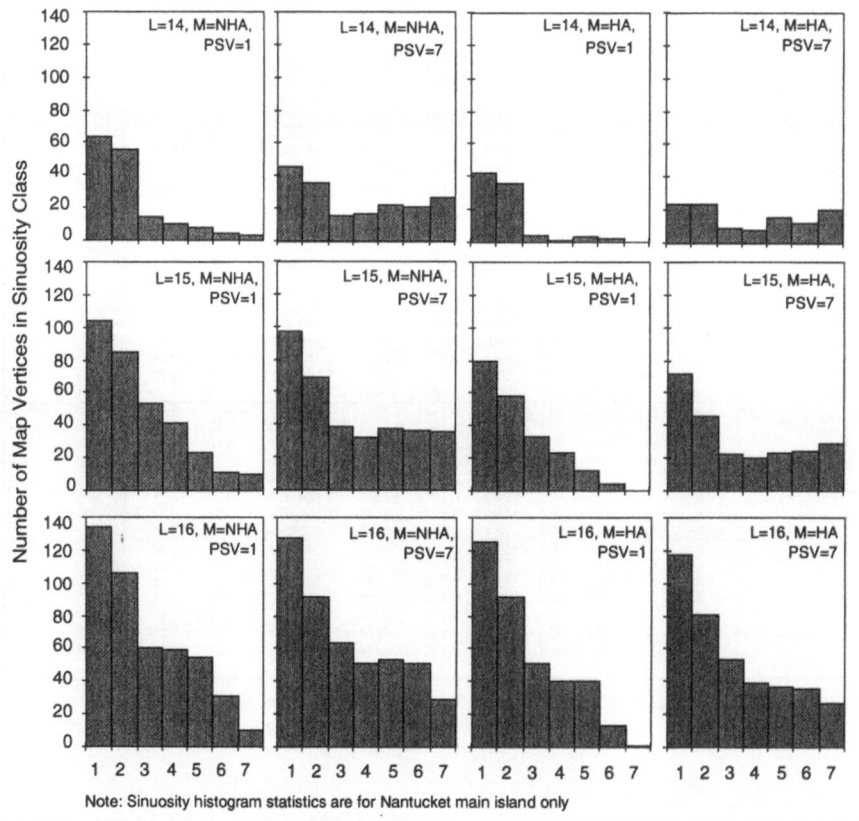

Note: Sinuosity histogram statistics are for Nantucket main island only

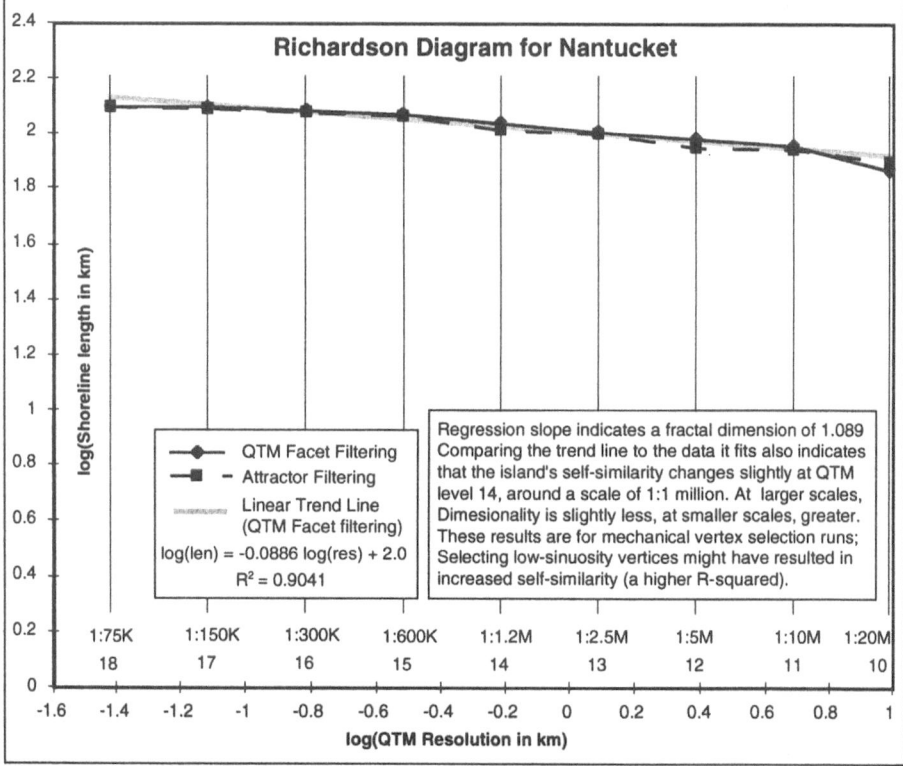

Figure C-2.3

Nantucket Island MA: Length Statistics for Main Island Coastline

Nantucket Island MA: Length Statistics for QTM Facet Filtering									
QTM LEVEL	18	17	16	15	14	13	12	11	10
VERTICES	820	759	628	459	274	164	90	53	31
LENGTH	125.6	124.7	122.5	117.6	109.8	102.1	95.9	90.9	74.2
SEG_LEN KM	0.153	0.165	0.195	0.257	0.402	0.626	1.078	1.748	2.473
QTM_RES KM	0.038	0.076	0.153	0.305	0.610	1.221	2.441	4.883	9.766
LOG(LENGTH)	2.099	2.096	2.088	2.070	2.041	2.009	1.982	1.959	1.870
LOG(SEG_LEN)	-0.814	-0.784	-0.709	-0.590	-0.396	-0.203	0.032	0.243	0.393
LOG(QTM_RES)	-1.419	-1.118	-0.816	-0.515	-0.214	0.087	0.388	0.689	0.990

Nantucket Island MA: Length Statistics for Attractor Filtering									
QTM LEVEL	18	17	16	15	14	13	12	11	10
VERTICES	783	617	437	311	141	84	42	25	19
LENGTH	125.2	123.3	120	116.7	104.2	100.1	88.9	88.1	78.9
SEG_LEN KM	0.160	0.200	0.275	0.376	0.744	1.206	2.168	3.671	4.383
QTM_RES KM	0.038	0.076	0.153	0.305	0.610	1.221	2.441	4.883	9.766
LOG(LENGTH)	2.098	2.091	2.079	2.067	2.018	2.000	1.949	1.945	1.897
LOG(SEG_LEN)	-0.796	-0.699	-0.560	-0.424	-0.128	0.081	0.336	0.565	0.642
LOG(QTM_RES)	-1.419	-1.118	-0.816	-0.515	-0.214	0.087	0.388	0.689	0.990

Richardson Diagram for Nantucket

Legend:
- QTM Facet Filtering
- Attractor Filtering
- Linear Trend Line (QTM Facet filtering)

log(len) = -0.0886 log(res) + 2.0

$R^2 = 0.9041$

Regression slope indicates a fractal dimension of 1.089 Comparing the trend line to the data it fits also indicates that the island's self-similarity changes slightly at QTM level 14, around a scale of 1:1 million. At larger scales, Dimesionality is slightly less, at smaller scales, greater. These results are for mechanical vertex selection runs; Selecting low-sinuosity vertices might have resulted in increased self-similarity (a higher R-squared).

y-axis: log(Shoreline length in km)

x-axis: log(QTM Resolution in km)

| 1:75K | 1:150K | 1:300K | 1:600K | 1:1.2M | 1:2.5M | 1:5M | 1:10M | 1:20M |
| 18 | 17 | 16 | 15 | 14 | 13 | 12 | 11 | 10 |

Appendix D

Generalization Experiments: Cartographic Results

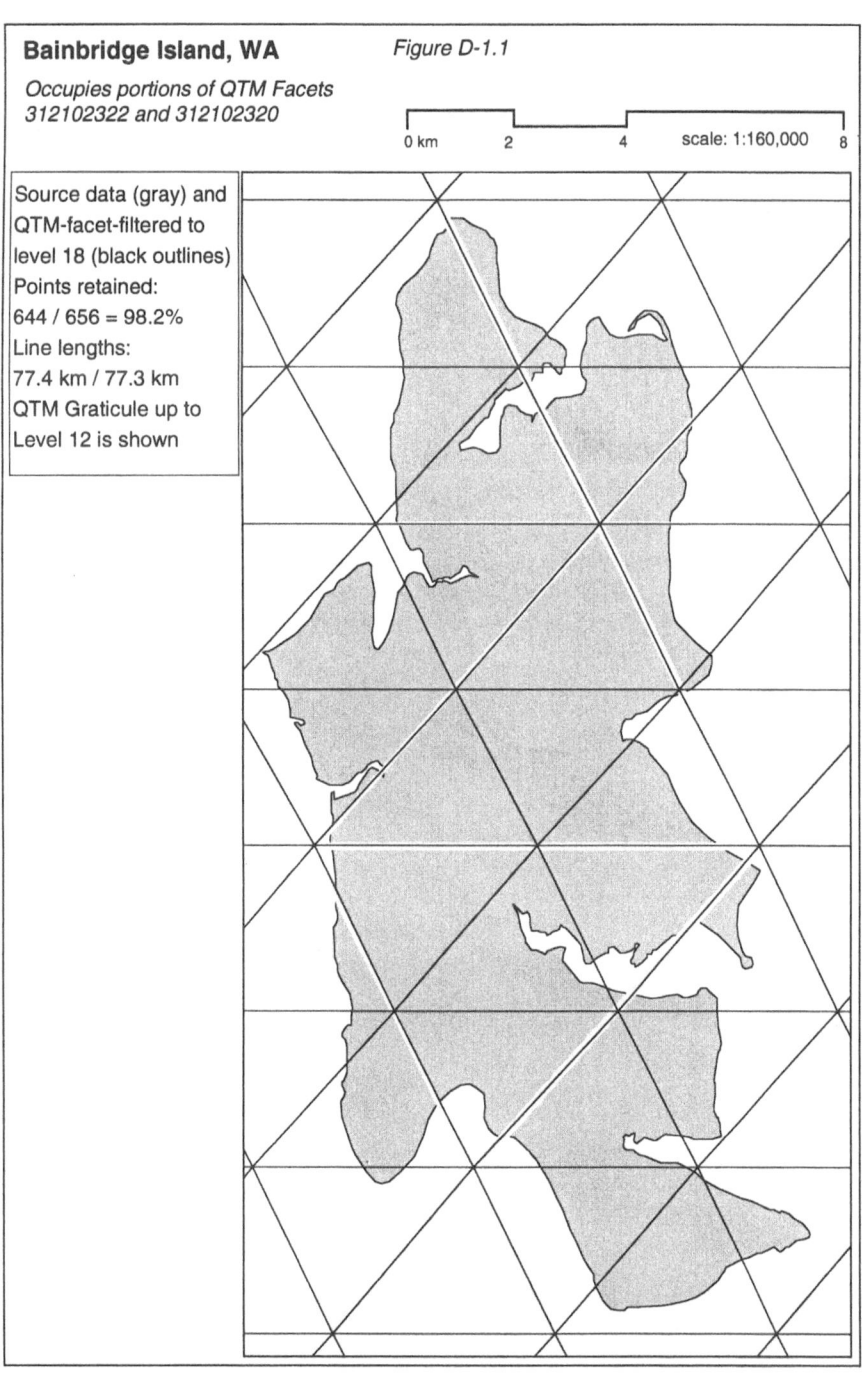

Bainbridge Island, WA   *Figure D-1.1*

Occupies portions of QTM Facets
312102322 and 312102320

0 km   2   4   scale: 1:160,000   8

Source data (gray) and
QTM-facet-filtered to
level 18 (black outlines)
Points retained:
644 / 656 = 98.2%
Line lengths:
77.4 km / 77.3 km
QTM Graticule up to
Level 12 is shown

Additional material from *A Hierarchical Coordinate System for Geoprocessing and Cartography*
ISBN 978-3-540-64980-9 (978-3-540-64980-9_OSFO1),
is available at http://extras.springer.com

Figure D-1.8

Bainbridge Island, WA: QTM Facet, Attractor and Bandwidth-tolerance Generalization Methods Compared

QTM ID filtering is a simple selection of one point from a set of contiguous ones occupying a QTM facet. Attractor filtering similarly selects from runs of points occupying an attractor (here using a one-point spike lookahead). Ramer-Douglas-Peucker (RDP) filtering uses a global bandwidth-tolerance operator, with tolerances selected to yield circa the same total number of points as the Attractor method at each given QTM level of detail.

Unfiltered 1:80K Source Data
656 points

0   5 km   10

Maps are plotted at 1:260,000. This is the nominal ("standard") scale of data filtered to QTM level 16 (top row). The QTM (areal) and RDP (linear) tolerances are shown alongside each map.

RDP filtering appears to have coarser results than the equivalent QTM Attractor filtering, even though about the same numbers of points are retained. One reason is that all QTM-based filters yield more evenly-spaced points than do band-tolerance methods, which tend to "iron out" stretches of low-sinuosity lines regardless of their extents. The RDP method also tends to be sensitive to its start/end points, which for polygons are usually arbitrarily chosen.

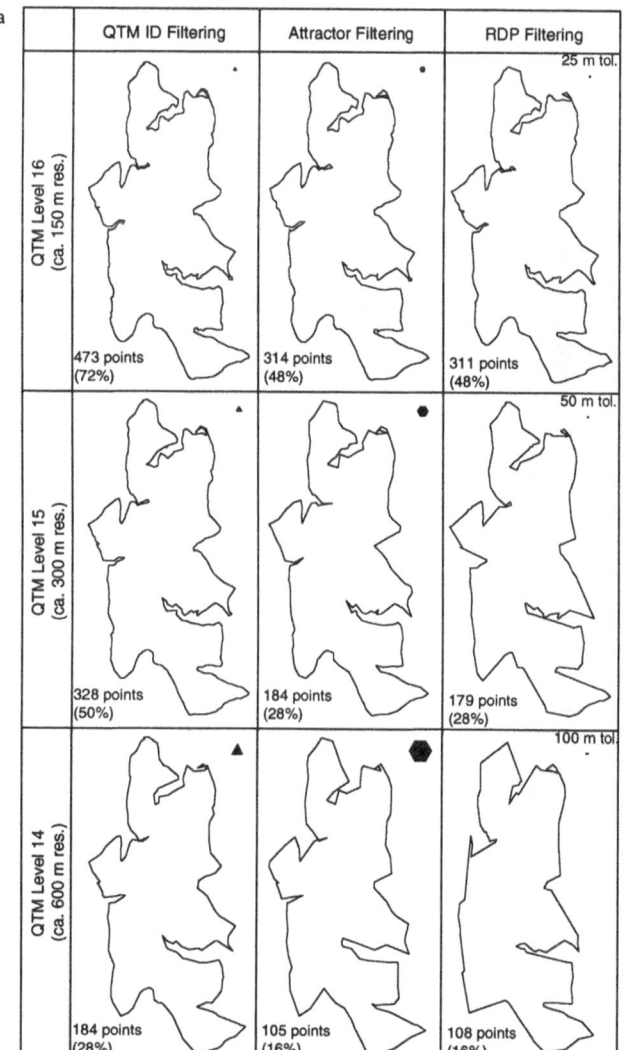

	QTM ID Filtering	Attractor Filtering	RDP Filtering
QTM Level 16 (ca. 150 m res.)	473 points (72%)	314 points (48%)	25 m tol. 311 points (48%)
QTM Level 15 (ca. 300 m res.)	328 points (50%)	184 points (28%)	50 m tol. 179 points (28%)
QTM Level 14 (ca. 600 m res.)	184 points (28%)	105 points (16%)	100 m tol. 108 points (16%)

Figure D-1.9

Bainbridge Island, WA: QTM Attractor Generalization
Compared with Source Data at Derived Scales

Constrained
generalization parameters:

Mode: *Non-hierarchical*
Mesh: *QTM attractors*
Lookahead: *1*
Points/Run: *1*
Sinuosity Lvl: *7 of 7*

NOAA Digital Chart data and World Vector Shorline data
were filtered to 250K and 1:200K to compare with
each other and U.S. CIA World Data Bank II data.
Compare same-scale maps along minor diagonals.
Tinted regions are footprints of source datasets.

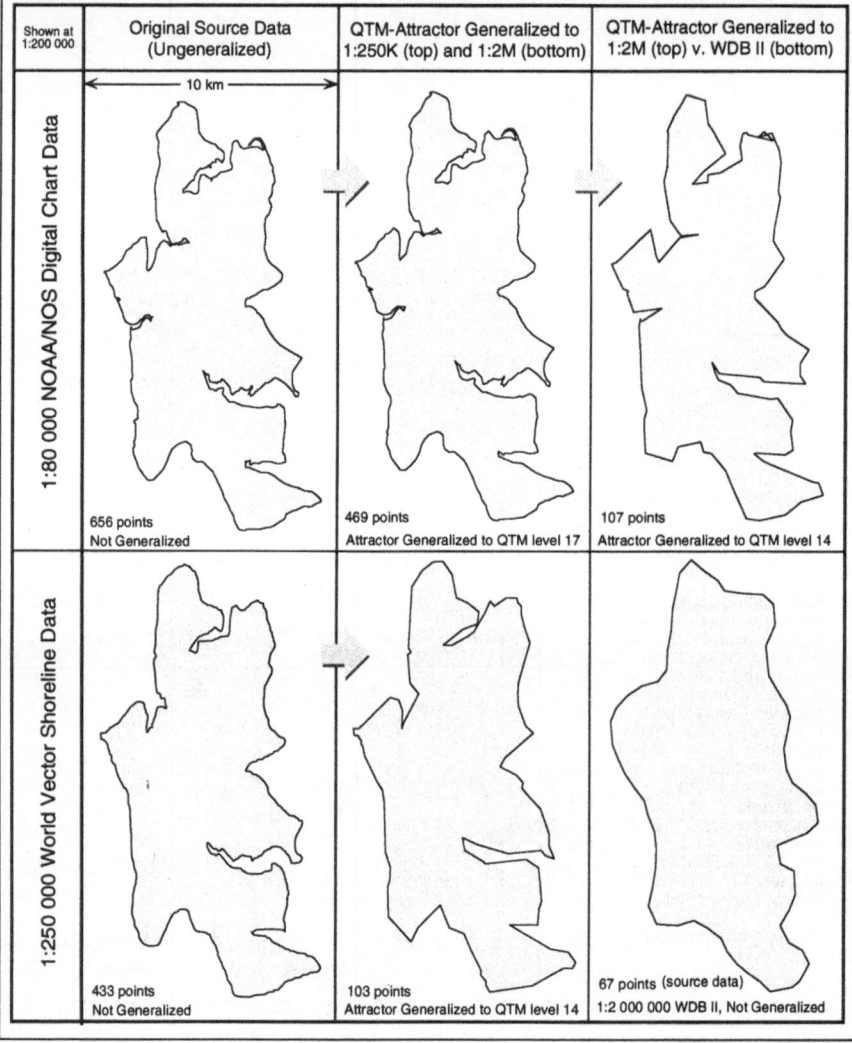

Shown at 1:200 000	Original Source Data (Ungeneralized)	QTM-Attractor Generalized to 1:250K (top) and 1:2M (bottom)	QTM-Attractor Generalized to 1:2M (top) v. WDB II (bottom)
1:80 000 NOAA/NOS Digital Chart Data	← 10 km → 656 points Not Generalized	469 points Attractor Generalized to QTM level 17	107 points Attractor Generalized to QTM level 14
1:250 000 World Vector Shoreline Data	433 points Not Generalized	103 points Attractor Generalized to QTM level 14	67 points (source data) 1:2 000 000 WDB II, Not Generalized

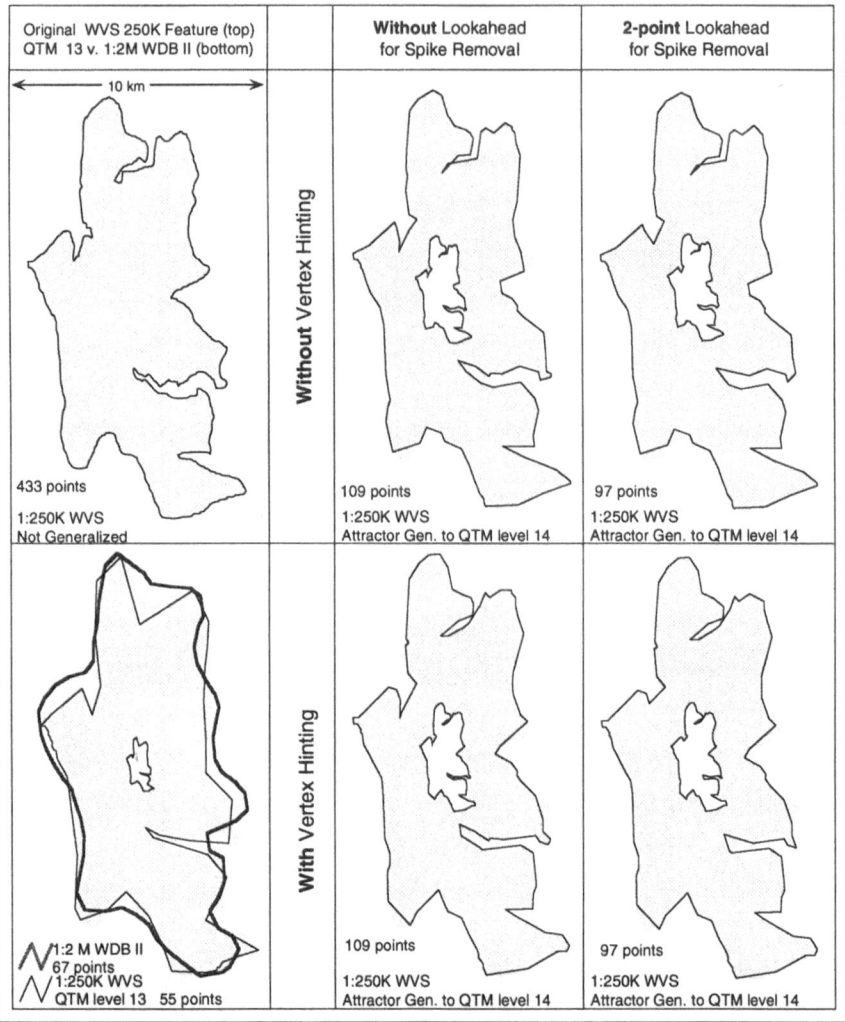

Figure D-1.10

Bainbridge Island, WA: Effects of QTM Lookahead and Sinuosity Parameters on World Vector Shoreline Data

Constrained
generalization parameters:

Mode: *Non-hierarchical*
Mesh: *QTM attractors*
Lookahead: *0 and 2*
Points/Run: *1*
Sinuosity Lvl: *0 and 7 of 7*

White insets
show simplified
maps at std. scale

The level 14 generalizations show minor variations caused by looking ahead for "spikes" and selecting vertices with high sinuosity values, or ignoring these.
A level 13 Attractor generalization (lower left) without lookahead or hints was made to compare with World Data Bank II data (ca. 1:2,000,000).

Original WVS 250K Feature (top) QTM 13 v. 1:2M WDB II (bottom)	**Without** Lookahead for Spike Removal	**2-point** Lookahead for Spike Removal
← 10 km →		
Without Vertex Hinting		
433 points 1:250K WVS Not Generalized	109 points 1:250K WVS Attractor Gen. to QTM level 14	97 points 1:250K WVS Attractor Gen. to QTM level 14
With Vertex Hinting		
1:2 M WDB II 67 points 1:250K WVS QTM level 13 55 points	109 points 1:250K WVS Attractor Gen. to QTM level 14	97 points 1:250K WVS Attractor Gen. to QTM level 14

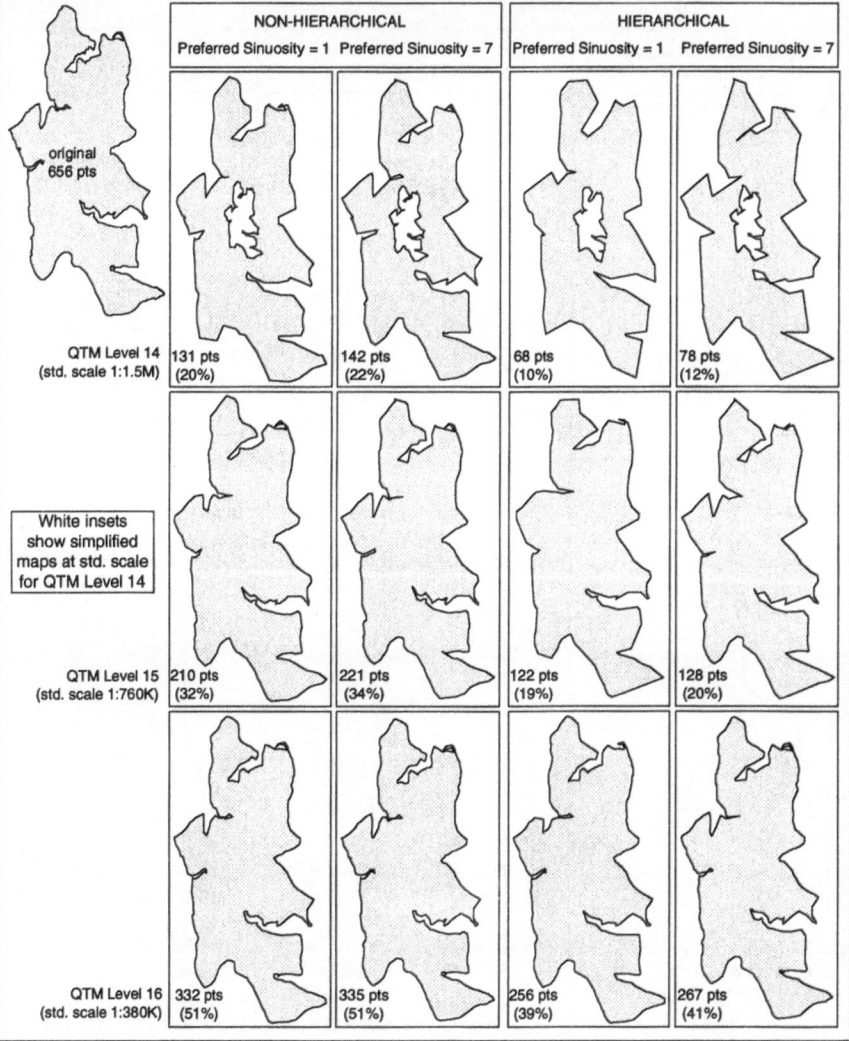

Figure D-1.11

Bainbridge Island, WA: Effects of Specifying Preferred Sinuosity for Non-hierarchical and Hierarchical QTM Line Simplification

0 km 2   4   8
Scale: 1:250,000
Mesh: *QTM Attractors*
Line Weight: *0.4 point*

Specifying a preferred sinuosity value (PSV) for vertices has little apparent effect at higher levels of detail, but can make a decided difference as scale ratios grow more extreme. Using a high PSV can somewhat compensate for the smoothing effect resulting from generalizing hierarchically, but low PSVs usually seem to work better.

	NON-HIERARCHICAL		HIERARCHICAL	
	Preferred Sinuosity = 1	Preferred Sinuosity = 7	Preferred Sinuosity = 1	Preferred Sinuosity = 7

original 656 pts

White insets show simplified maps at std. scale for QTM Level 14

QTM Level 14 (std. scale 1:1.5M): 131 pts (20%) · 142 pts (22%) · 68 pts (10%) · 78 pts (12%)

QTM Level 15 (std. scale 1:760K): 210 pts (32%) · 221 pts (34%) · 122 pts (19%) · 128 pts (20%)

QTM Level 16 (std. scale 1:380K): 332 pts (51%) · 335 pts (51%) · 256 pts (39%) · 267 pts (41%)

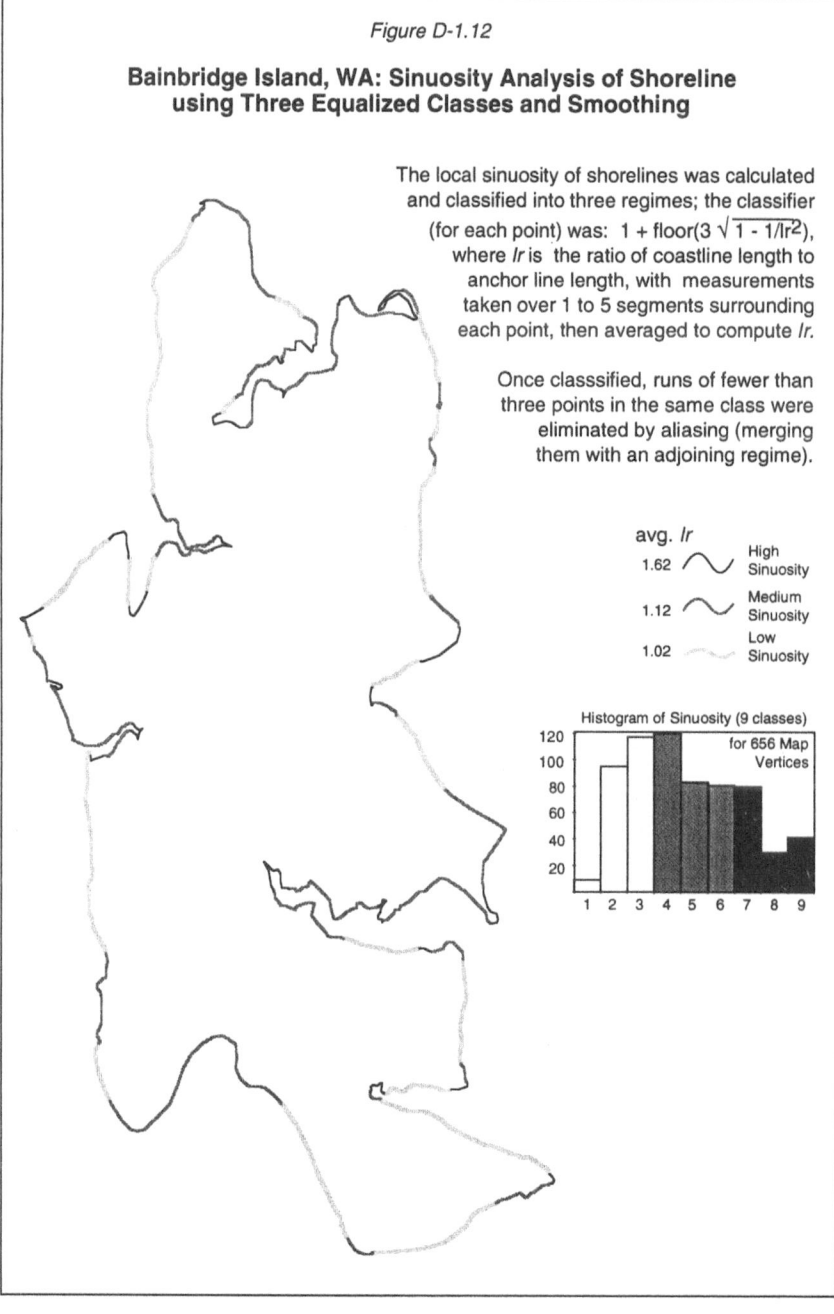

Figure D-1.12

Bainbridge Island, WA: Sinuosity Analysis of Shoreline using Three Equalized Classes and Smoothing

198

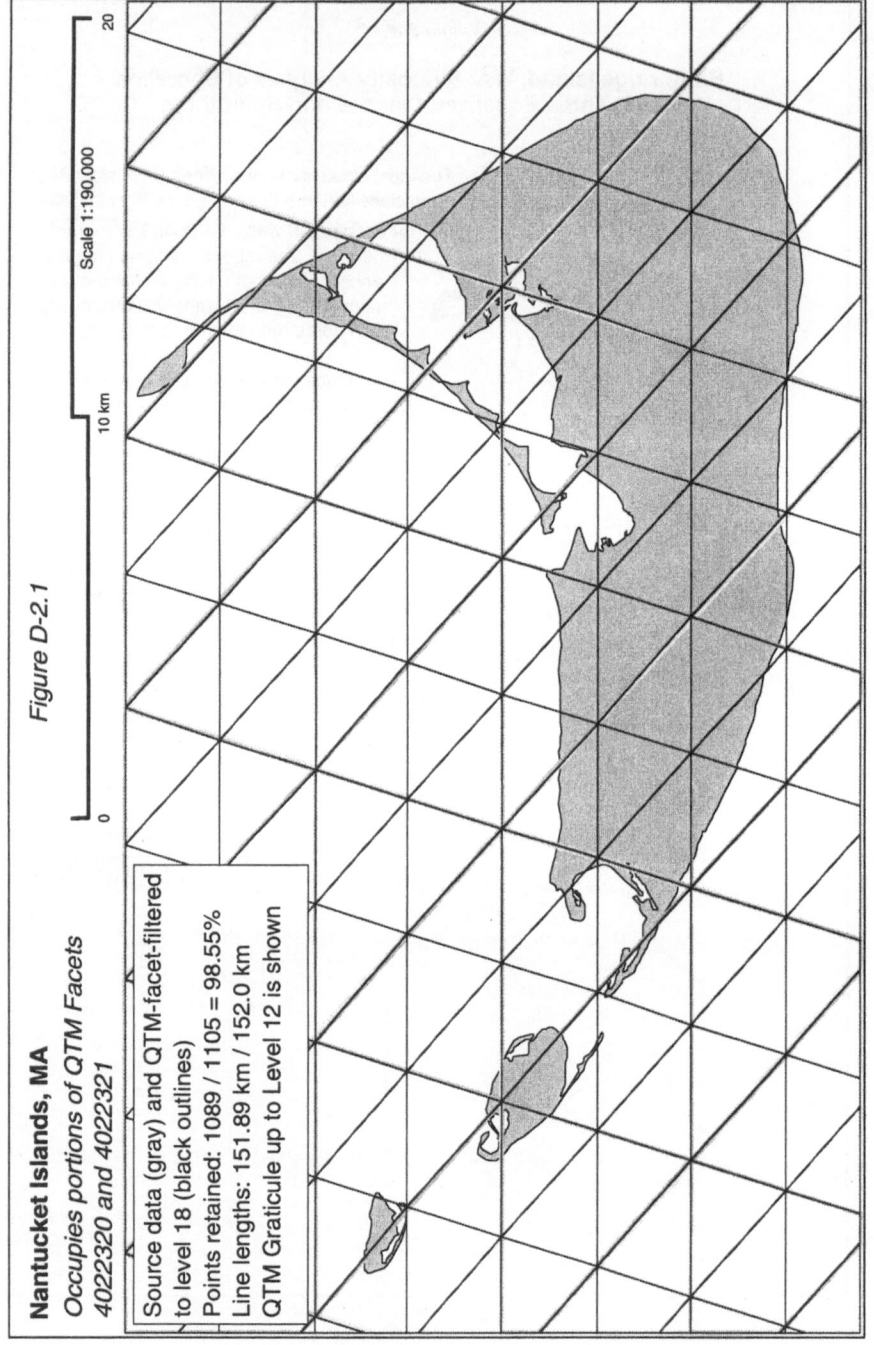

Nantucket Islands, MA

*Occupies portions of QTM Facets
4022320 and 4022321*

Source data (gray) and QTM-facet-filtered
to level 18 (black outlines)

Points retained: 1089 / 1105 = 98.55%

Line lengths: 151.89 km / 152.0 km

QTM Graticule up to Level 12 is shown

Figure D-2.1

Scale 1:190,000

0   10 km   20

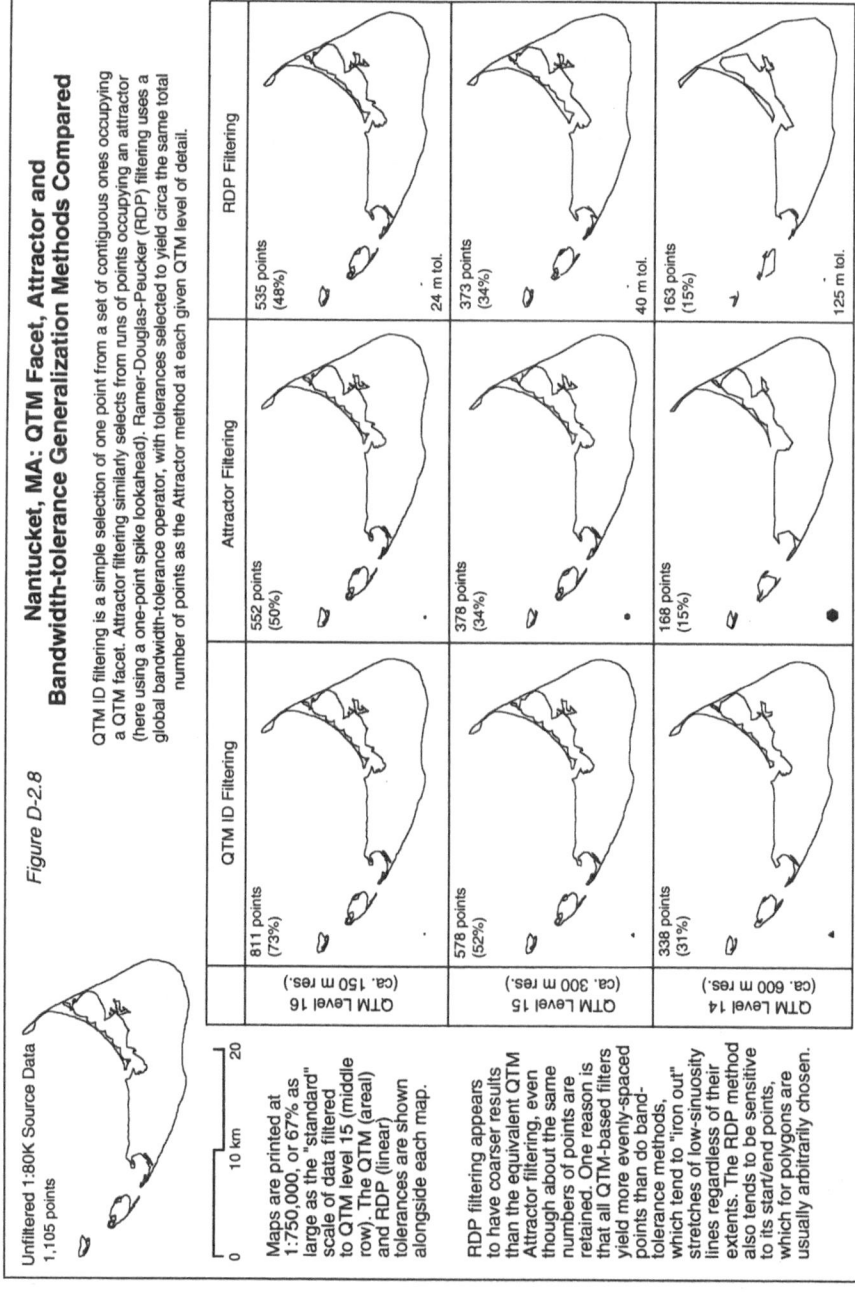

Figure D-2.8

Nantucket, MA: QTM Facet, Attractor and Bandwidth-tolerance Generalization Methods Compared

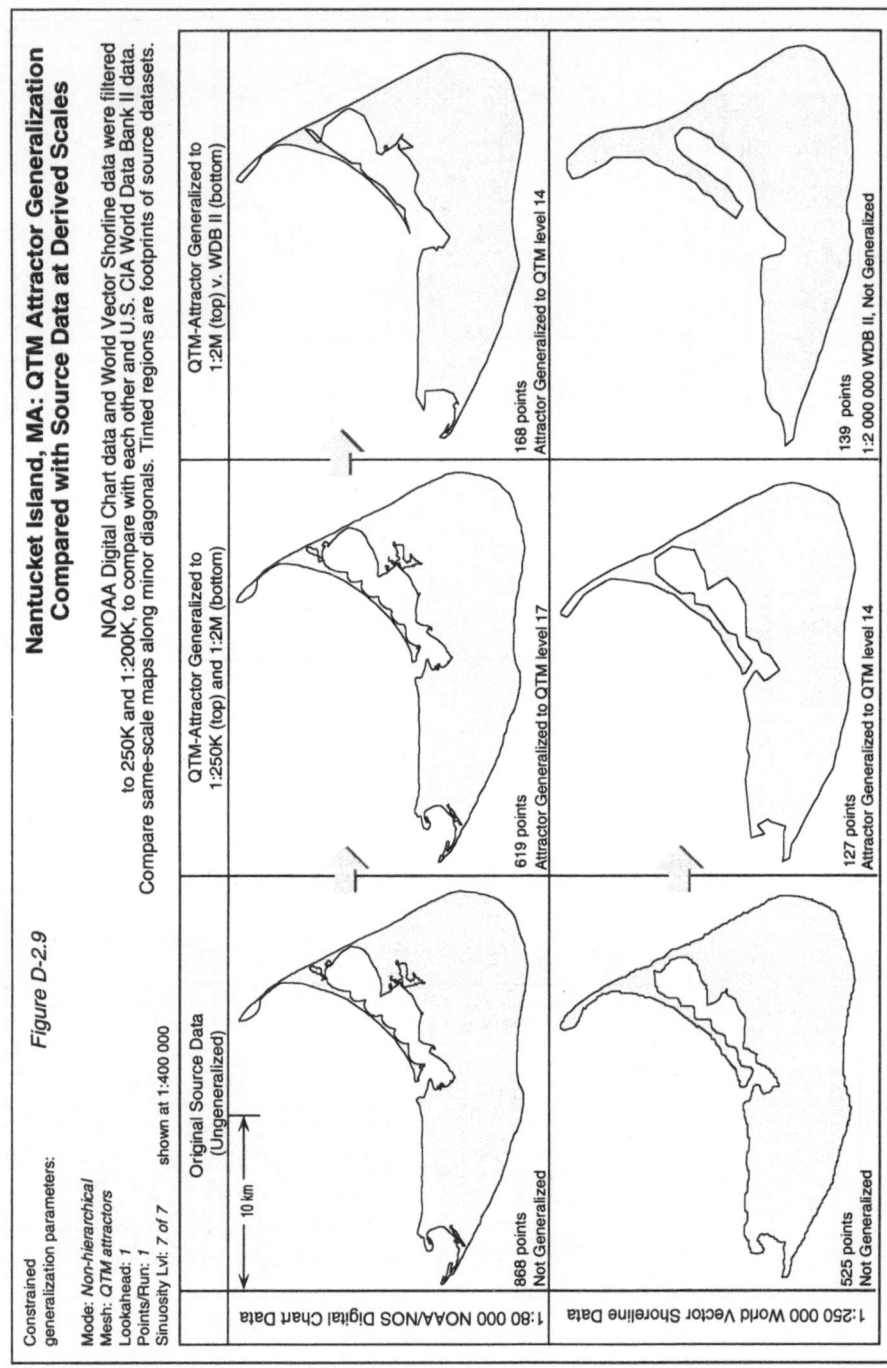

Figure D-2.9

Nantucket Island, MA: QTM Attractor Generalization
Compared with Source Data at Derived Scales

NOAA Digital Chart data and World Vector Shoreline data were filtered to 250K and 1:200K, to compare with each other and U.S. CIA World Data Bank II data. Compare same-scale maps along minor diagonals. Tinted regions are footprints of source datasets.

Constrained generalization parameters:

Mode: *Non-hierarchical*
Mesh: *QTM attractors*
Lookahead: *1*
Points/Run: *1*
Sinuosity Lvl: *7 of 7*   shown at 1:400 000

Original Source Data (Ungeneralized)

10 km

1:80 000 NOAA/NOS Digital Chart Data

868 points
Not Generalized

QTM-Attractor Generalized to 1:250K (top) and 1:2M (bottom)

619 points
Attractor Generalized to QTM level 17

QTM-Attractor Generalized to 1:2M (top) v. WDB II (bottom)

168 points
Attractor Generalized to QTM level 14

1:250 000 World Vector Shoreline Data

525 points
Not Generalized

127 points
Attractor Generalized to QTM level 14

139 points
1:2 000 000 WDB II, Not Generalized

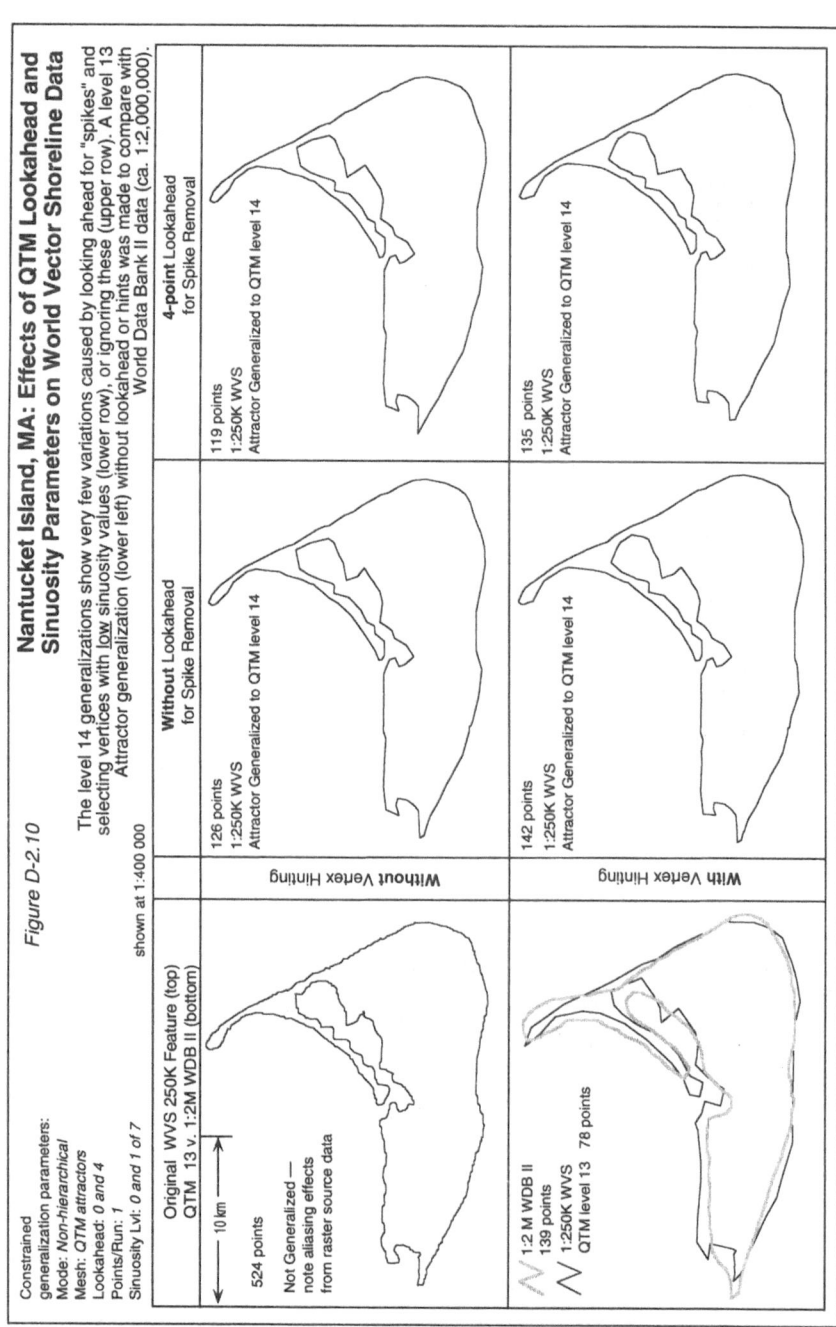

Figure D-2.10

Nantucket Island, MA: Effects of QTM Lookahead and Sinuosity Parameters on World Vector Shoreline Data

208

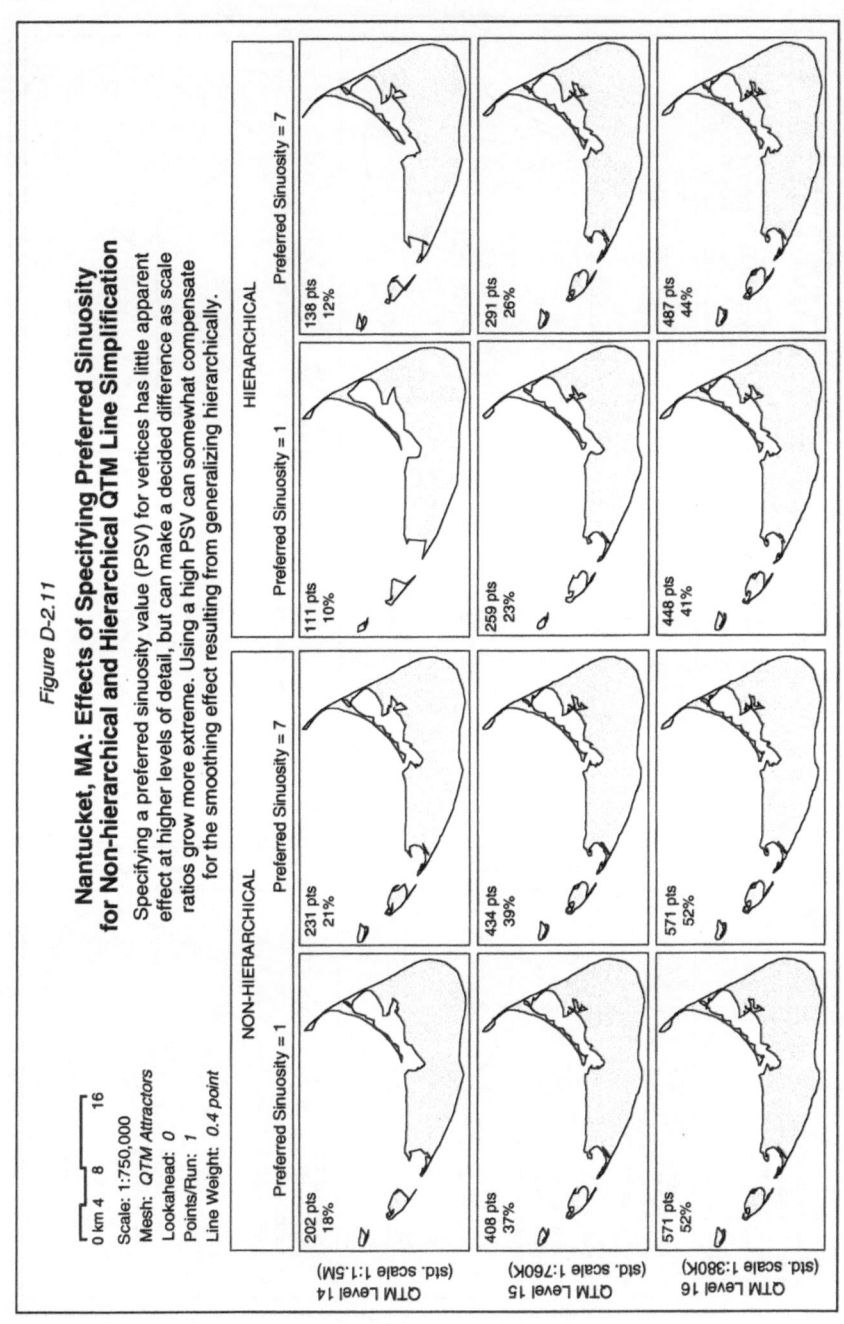

Figure D-2.11

**Nantucket, MA: Effects of Specifying Preferred Sinuosity
for Non-hierarchical and Hierarchical QTM Line Simplification**

Specifying a preferred sinuosity value (PSV) for vertices has little apparent effect at higher levels of detail, but can make a decided difference as scale ratios grow more extreme. Using a high PSV can somewhat compensate for the smoothing effect resulting from generalizing hierarchically.

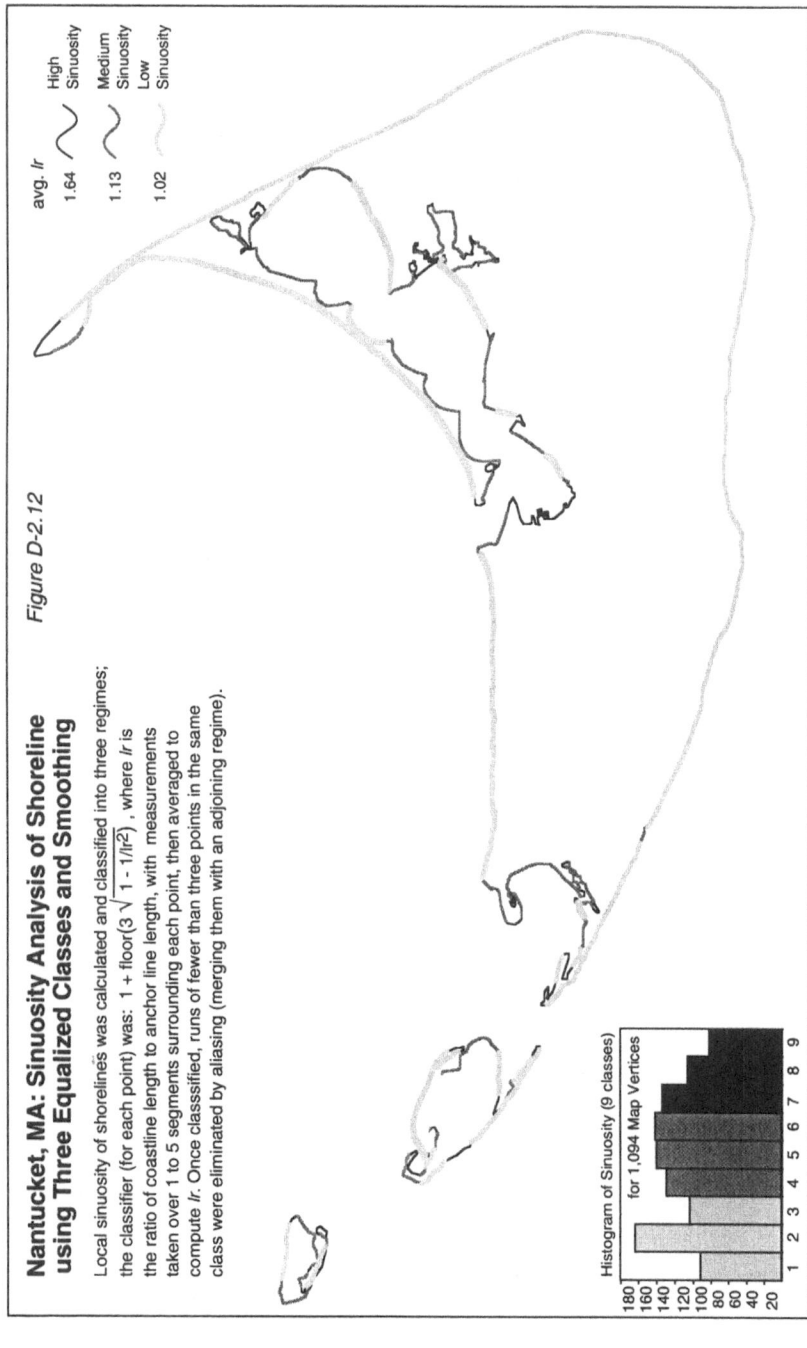

Nantucket, MA: Sinuosity Analysis of Shoreline using Three Equalized Classes and Smoothing

Figure D-2.12

Local sinuosity of shorelines was calculated and classified into three regimes; the classifier (for each point) was: $1 + \text{floor}(3\sqrt{1 - 1/lr^2})$, where lr is the ratio of coastline length to anchor line length, with measurements taken over 1 to 5 segments surrounding each point, then averaged to compute lr. Once classsified, runs of fewer than three points in the same class were eliminated by aliasing (merging them with an adjoining regime).

avg. lr		
1.64	$\sim$	High Sinuosity
1.13	$\sim$	Medium Sinuosity
1.02	$\cdots$	Low Sinuosity

Histogram of Sinuosity (9 classes)
for 1,094 Map Vertices

Appendix E

Efficiency of QTM as a Spatial Index for Point Data

To determine how efficiently QTM encoding can be for indexing spatial data, point locations for 60 cities and towns scattered across the globe were assembled from a gazetteer. An unlikely collection of places was chosen to insure that when ordered by city name, successive cities are not located nearby, so that a linear spatial search of this list would be highly inefficient. Three indexed orderings were then generated: by *latitudes*, by *longitudes* and by *QTM identifiers*. All of these are monotonic spatial orderings, the first two being one-dimensional indices, the last being a two-dimensional index.

As the input coordinates were accurate only to one arc minute, QTM Identifiers were computed to 12 levels of detail, roughly a 1 to 2 km resolution. The closest pair of cities are 1.1 km apart; their QTM IDs differ only in the tenth digit. Most other cities are several hundred and even thousands of kilometers from their closest neighbor, complicating spatial indexing. Ordered by name, list neighbors are an average of 7,722 km apart, with a standard deviation of 4,540 km.

Great circle distances were computed between adjacent list items and also cumulated to obtain the total distance that one would travel in flying from the first to the last city on the lists and then to the starting point, visiting all intermediate citys in order. These "tours", which represent a solution to the "travelling salesman" problem for each of the orderings, are shown in tables E1-E6, and graphed in Figures E7 and E8. The figures and charts indicate that that indexing performance is best for QTM IDs, next best for Longitudes, then Latitudes, and poorest for City Name.

To provide yardsticks with which to evaluate indexing efficiency, a shortest path algorithm (a greedy, approximate one) was applied to these locations to yield a best case indexing based on total tour distance (other criteria obviously would have to be considered in actual GIS applications). The algorithm was then modified to approximate the longest possible path, and these statistics were used to standardize the results described above. These best- and worst-case solutions are tabulated in E-5 and E-6, respectively. Figure E7 displays distance legs from E1-E6 as line graphs, and E-8 graphs cumulative distances above and bar charts below, which compares total distance for each method, standardized by shortest/longest path lengths. QTM indexing is found to be 4 times as efficient as the worst case (longest path), and 87% as good as the best case (shortest path).

E 1:   Spatial Ordering Statistics for 60 Cities and Towns, ordered by City Name

E 2:   Spatial Ordering Statistics for 60 Cities and Towns, ordered by Latitude

E 3:   Spatial Ordering Statistics for 60 Cities and Towns, ordered by Longitude

E 4:   Spatial Ordering Statistics for 60 Cities and Towns, ordered by QTM ID

E 5:   Spatial Ordering Statistics for 60 Cities and Towns, ordered by Shortest Path

E 6:   Spatial Ordering Statistics for 60 Cities and Towns, ordered by Longest Path

E 7:   Distance from Prior City on List, by Type of Ordering

E 8:   Cumulative Distance for City Tours and Efficiency Scores

ORD	CITY	COUNTRY	STATE/PROV	LON DD	LAT DD	QTM ID	DIST LAST	DIST CUM
Table E-1					MEAN KM:	7721.9		
Spatial Ordering Statistics for 60 Cities and Towns					STD. DEV. KM:	4539.8		
Sorted by:	CITY NAME				MIN & MAX KM:	1.1	17101.1	
1	Athens	Greece		23.72	37.97	1030030332222	1617.6	1617.6
2	Athens	US	Maine	-69.67	44.95	4022333113320	7368.6	8986.1
3	Berlin	Germany		13.37	52.50	1130131231000	5801.6	14787.8
4	Berlin	South Africa		27.58	-32.90	5032022030212	9588.7	24376.5
5	Bombay	India		72.83	18.97	1300201313222	7501.7	31878.2
6	Bombay	US	Minnesota	-92.98	44.27	3022220022010	12826.9	44705.1
7	Boston	England		-0.02	52.98	4133123321322	6405.7	51110.8
8	Boston	Philippines		126.37	-7.87	6310213030303	13065.8	64176.5
9	Boston	US	Massachusetts	-71.07	42.37	4022301203222	15803.4	79980.0
10	Cairo	Egypt		31.25	30.05	1032313200120	8711.5	88691.5
11	Cairo	US	Illinois	-88.82	37.00	4322312021122	10286.7	98978.2
12	Colombo	Brazil		-49.23	-25.28	8003101331202	8063.6	107041.7
13	Colombo	Sri Lanka		79.85	6.93	1330020130001	14236.2	121277.9
14	Dutton	England		-2.63	53.32	4133103103210	8884.2	130162.2
15	Dutton	US	Virginia	-76.47	37.50	4022132001302	5741.7	135903.9
16	Fairbanks	US	Alaska	-147.72	64.85	3102101033201	5416.0	141319.8
17	Fairbanks	US	Florida	-82.27	29.73	4320310200030	5885.5	147205.4
18	Jerusalem	Israel		35.23	31.77	1002332101002	5112.9	152318.3
19	Jerusalem	US	Arkansas	-92.82	35.40	3322300303130	10779.4	163097.7
20	Johannesburg	South Africa		28.00	-26.25	5032201301200	14345.2	177442.9
21	Johannisburg	Germany		21.82	53.62	1130303323221	8894.4	186337.4
22	Johannisburg	Poland		21.82	53.63	1130303323201	1.1	186338.5
23	Lagos	Angola		17.05	-16.07	5200013122120	7757.8	194096.3
24	Lagos	Nigeria		3.40	6.45	1222323202312	2918.0	197014.4
25	Lagos	Portugal		-8.67	37.10	4233021133111	3618.4	200632.7
26	Los Angeles	Chile		-72.35	-37.47	8022103213202	10550.0	211182.8
27	Los Angeles	US	California	-118.25	34.07	3020233012200	9255.2	220438.0
28	Los Angeles	US	Texas	-99.00	28.47	3320011020222	1927.5	222365.5
29	Moscow	Russia		37.58	55.75	1101302200021	9779.1	232144.6
30	Moscow	US	Mississippi	-88.65	32.77	4321333120230	8927.2	241071.9
31	Paris	France		2.33	48.87	1133320101111	7392.5	248464.3
32	Paris	US	Texas	-95.55	33.67	3320321122013	7774.3	256238.6
33	Rome	Italy		12.48	41.90	1033003121132	8857.4	265095.9
34	Rome	US	Ohio	-82.42	38.47	4322030320121	7616.7	272712.6
35	Santiago	Bolivia		-59.57	-18.32	8313300201112	6744.4	279457.1
36	Santiago	Brazil		-54.88	-29.18	8003210103222	1297.1	280754.2
37	Santiago	Chile		-70.67	-33.45	8023111023230	1570.2	282324.4
38	Santiago	Costa Rica		-84.30	9.85	4330210322333	5022.3	287346.7
39	Santiago	Cuba		-75.82	20.02	4301223332310	1450.2	288796.9
40	Santiago	Dominican Rep.		-70.70	19.45	4303231303213	539.2	289336.1
41	Santiago	Mexico		-99.85	25.42	3321213111303	3059.6	292395.7
42	Santiago	Mexico		-109.72	23.47	3323210132022	1021.3	293417.1
43	Santiago	Panama		-80.98	8.10	4330003303103	3503.4	296920.4
44	Santiago	Paraguay		-56.78	-27.15	8003311032233	4703.8	301624.3
45	Santiago	Peru		-75.73	-14.18	8300312130230	2435.1	304059.4
46	St.Petersburg	Russia		30.25	59.92	1102301202022	12247.6	316307.0
47	St.Petersburg	US	Pennsylvania	-79.65	41.17	4022232132221	7090.9	323397.9
48	Stockholm	Sweden		18.05	59.33	1132313023233	6557.6	329955.5
49	Stockholm	US	Maine	-68.13	47.05	4120330022123	5472.6	335428.0
50	Sydney	Australia		151.22	-33.87	6030322332302	16412.7	351840.8
51	Sydney	CAN	Nova Scotia	-60.18	46.15	4121320200031	17026.4	368867.2
52	Wellington	Australia		148.95	-32.55	6032301221221	17101.1	385968.2
53	Wellington	CAN	Ontario	-77.35	43.95	4022011232332	15826.3	401794.5
54	Wellington	New Zealand		174.78	-41.30	6233321230222	14290.1	416084.7
55	Wellington	South Africa		18.95	-33.63	5032111130232	11315.4	427400.1
56	Zurich	CAN	Ontario	-81.62	43.43	4022202111312	13272.2	440672.3
57	Zurich	Netherlands		5.38	53.10	1133102031320	6118.9	446791.1
58	Zurich	US	Kansas	-99.43	39.23	3322033301212	7471.2	454262.3
59	Zurich	US	Montana	-109.03	48.58	3120301202200	1290.1	455552.4
60	Zürich	Switzerland		8.53	47.38	1133013131311	7760.7	463313.1

Table E-2						MEAN KM:	2734.5	
Spatial Ordering Statistics for 60 Cities and Towns						STD. DEV. KM:	3726.4	
Sorted by:	QTM ID					MIN & MAX KM	1.1	16286.0
ORD	CITY	COUNTRY	STATE/PROV	LON DD	LAT DD	QTM ID	DIST LAST	DIST CUM
1	Jerusalem	Israel		35.23	31.77	1002332101002	11497.3	11497.3
2	Athens	Greece		23.72	37.97	1030030332222	4866.9	16364.2
3	Cairo	Egypt		31.25	30.05	1032313200120	1119.5	17483.7
4	Rome	Italy		12.48	41.90	1033003121132	2132.4	19616.1
5	Moscow	Russia		37.58	55.75	1101302200021	2372.3	21988.4
6	St.Petersburg	Russia		30.25	59.92	1102301202022	633.9	22622.3
7	Berlin	Germany		13.37	52.50	1130131231000	1323.1	23945.4
8	Johannisburg	Poland		21.82	53.63	1130303323201	577.6	24523.0
9	Johannisburg	Germany		21.82	53.62	1130303323221	1.1	24524.1
10	Stockholm	Sweden		18.05	59.33	1132313023233	675.0	25199.1
11	Zürich	Switzerland		8.53	47.38	1133013131311	1466.9	26666.0
12	Zurich	Netherlands		5.38	53.10	1133102031320	673.7	27339.7
13	Paris	France		2.33	48.87	1133320101111	516.0	27855.7
14	Lagos	Nigeria		3.40	6.45	1222323202312	4714.4	32570.1
15	Bombay	India		72.83	18.97	1300201313222	7609.9	40180.0
16	Colombo	Sri Lanka		79.85	6.93	1330020130001	1537.9	41717.8
17	Los Angeles	US	California	-118.25	34.07	3020233012200	15062.5	56780.4
18	Bombay	US	Minnesota	-92.98	44.27	3022220022010	2440.2	59220.5
19	Fairbanks	US	Alaska	-147.72	64.85	3102101033201	4015.9	63236.5
20	Zurich	US	Montana	-109.03	48.58	3120301202200	2896.8	66133.3
21	Los Angeles	US	Texas	-99.00	28.47	3320011020222	2393.6	68526.9
22	Paris	US	Texas	-95.55	33.67	3320321122013	664.4	69191.3
23	Santiago	Mexico		-99.85	25.42	3321213111303	1006.2	70197.5
24	Zurich	US	Kansas	-99.43	39.23	3322033301212	1534.9	71732.5
25	Jerusalem	US	Arkansas	-92.82	35.40	3322300303130	722.4	72454.8
26	Santiago	Mexico		-109.72	23.47	3323210132022	2099.5	74554.3
27	Wellington	CAN	Ontario	-77.35	43.95	4022011232332	3718.9	78273.2
28	Dutton	US	Virginia	-76.47	37.50	4022132001302	720.5	78993.7
29	Zurich	CAN	Ontario	-81.62	43.43	4022202111312	789.3	79783.0
30	St.Petersburg	US	Pennsylvania	-79.65	41.17	4022232132221	298.8	80081.8
31	Boston	US	Massachusetts	-71.07	42.37	4022301203222	723.1	80804.8
32	Athens	US	Maine	-69.67	44.95	4022333113320	308.0	81112.8
33	Stockholm	US	Maine	-68.13	47.05	4120330022123	261.9	81374.6
34	Sydney	CAN	Nova Scotia	-60.18	46.15	4121320200031	614.8	81989.5
35	Dutton	England		-2.63	53.32	4133103103210	4092.4	86081.9
36	Boston	England		-0.02	52.98	4133123321322	178.0	86259.9
37	Lagos	Portugal		-8.67	37.10	4233021133111	1887.4	88147.2
38	Santiago	Cuba		-75.82	20.02	4301233332310	6685.0	94832.2
39	Santiago	Dominican Rep.		-70.70	19.45	4303231303213	539.2	95371.4
40	Fairbanks	US	Florida	-82.27	29.73	4320310200030	1632.3	97003.7
41	Moscow	US	Mississippi	-88.65	32.77	4321333120230	693.6	97697.3
42	Rome	US	Ohio	-82.42	38.47	4322030320121	846.7	98544.0
43	Cairo	US	Illinois	-88.82	37.00	4322312021122	585.5	99129.5
44	Santiago	Panama		-80.98	8.10	4330003303103	3307.2	102436.6
45	Santiago	Costa Rica		-84.30	9.85	4330210322333	413.0	102849.6
46	Berlin	South Africa		27.58	-32.90	5032022030212	12628.2	115477.8
47	Wellington	South Africa		18.95	-33.63	5032111130232	805.6	116283.4
48	Johannesburg	South Africa		28.00	-26.25	5032201301200	1195.5	117478.9
49	Lagos	Angola		17.05	-16.07	5200013122120	1600.4	119079.3
50	Sydney	Australia		151.22	-33.87	6030322332302	12631.4	131710.7
51	Wellington	Australia		148.95	-32.55	6032301221221	257.0	131967.7
52	Wellington	New Zealand		174.78	-41.30	6233321230222	2479.7	134447.4
53	Boston	Philippines		126.37	-7.87	6310213030303	6026.9	140474.4
54	Colombo	Brazil		-49.23	-25.28	8003101331202	16286.0	156760.4
55	Santiago	Brazil		-54.88	-29.18	8003210103222	706.5	157466.9
56	Santiago	Paraguay		-56.78	-27.15	8003311032233	292.4	157759.3
57	Los Angeles	Chile		-72.35	-37.47	8022103213202	1853.7	159613.0
58	Santiago	Chile		-70.67	-33.45	8023111023230	471.8	160084.8
59	Santiago	Peru		-75.73	-14.18	8300312130230	2201.1	162285.9
60	Santiago	Bolivia		-59.57	-18.32	8313300201112	1783.3	164069.2

Table E-3					MEAN KM:	5439.4	
Spatial Ordering Statistics for 60 Cities and Towns					STD. DEV. KM:	4928.5	
Sorted by:	LATITUDE				MIN & MAX KM:	1.1	17298.3
ORD	CITY	COUNTRY	STATE/PROV	LON DD	LAT DD	QTM ID	DIST LAST	DIST CUM
1	Wellington	New Zealand		174.78	-41.30	6233321230222	12236.5	12236.5
2	Los Angeles	Chile		-72.35	-37.47	8022103213202	8913.9	21150.4
3	Sydney	Australia		151.22	-33.87	6030322332302	10884.1	32034.5
4	Wellington	South Africa		18.95	-33.63	5032111130232	10999.5	43034.0
5	Santiago	Chile		-70.67	-33.45	8023111023230	7994.2	51028.2
6	Berlin	South Africa		27.58	-32.90	5032022030212	8725.4	59753.6
7	Wellington	Australia		148.95	-32.55	6032301221221	10485.4	70239.0
8	Santiago	Brazil		-54.88	-29.18	8003210103222	12695.6	82934.6
9	Santiago	Paraguay		-56.78	-27.15	8003311032233	292.4	83227.0
10	Johannesburg	South Africa		28.00	-26.25	5032201301200	8230.2	91457.2
11	Colombo	Brazil		-49.23	-25.28	8003101331202	7599.9	99057.2
12	Santiago	Bolivia		-59.57	-18.32	8313300201112	1316.6	100373.8
13	Lagos	Angola		17.05	-16.07	5200013122120	8072.9	108446.7
14	Santiago	Peru		-75.73	-14.18	8300312130230	9856.0	118302.6
15	Boston	Philippines		126.37	-7.87	6310213030303	16544.8	134847.4
16	Lagos	Nigeria		3.40	6.45	1222323202312	13715.4	148562.8
17	Colombo	Sri Lanka		79.85	6.93	1330020130001	8426.4	156989.3
18	Santiago	Panama		-80.98	8.10	4330003303103	17298.3	174287.6
19	Santiago	Costa Rica		-84.30	9.85	4330210322333	413.0	174700.6
20	Bombay	India		72.83	18.97	1300201313222	15934.1	190634.7
21	Santiago	Dominican Rep.		-70.70	19.45	4303231303213	14167.4	204802.1
22	Santiago	Cuba		-75.82	20.02	4301223332310	539.2	205341.3
23	Santiago	Mexico		-109.72	23.47	3232310132022	3511.6	208852.9
24	Santiago	Mexico		-99.85	25.42	3321213111303	1021.3	209874.2
25	Los Angeles	US	Texas	-99.00	28.47	3320011020222	349.2	210223.4
26	Fairbanks	US	Florida	-82.27	29.73	4320310200030	1628.8	211852.2
27	Cairo	Egypt		31.25	30.05	1032313200120	10328.8	222181.1
28	Jerusalem	Israel		35.23	31.77	1002332101002	5112.9	227294.0
29	Moscow	US	Mississippi	-88.65	32.77	4321333120230	10724.3	238018.2
30	Paris	US	Texas	-95.55	33.67	3320321122013	649.0	238667.2
31	Los Angeles	US	California	-118.25	34.07	3022300303130	2090.4	240757.6
32	Jerusalem	US	Arkansas	-92.82	35.40	3322300303130	2320.3	243077.9
33	Cairo	US	Illinois	-88.82	37.00	4322312021122	400.2	243478.2
34	Lagos	Portugal		-8.67	37.10	4233021133111	6870.7	250348.9
35	Dutton	US	Virginia	-76.47	37.50	4022132001302	5853.1	256202.0
36	Athens	Greece		23.72	37.97	1030030332222	4866.9	261068.9
37	Rome	US	Ohio	-82.42	38.47	4322030320121	8645.4	269714.4
38	Zurich	US	Kansas	-99.43	39.23	3322033301212	1472.2	271186.5
39	St.Petersburg	US	Pennsylvania	-79.65	41.17	4022232132221	1688.7	272875.2
40	Rome	Italy		12.48	41.90	1033003121132	7249.0	280124.2
41	Boston	US	Massachusetts	-71.07	42.37	4022301203222	6579.3	286703.5
42	Zurich	CAN	Ontario	-81.62	43.43	4022202111312	866.1	287569.7
43	Wellington	CAN	Ontario	-77.35	43.95	4022011232332	347.9	287917.5
44	Bombay	US	Minnesota	-92.98	44.27	3022220022010	1245.6	289163.1
45	Athens	US	Maine	-69.67	44.95	4022333113320	1839.0	291002.1
46	Sydney	CAN	Nova Scotia	-60.18	46.15	4121320200031	749.9	291752.0
47	Stockholm	US	Maine	-68.13	47.05	4120330022123	614.8	292366.8
48	Zürich	Switzerland		8.53	47.38	1133013131311	5536.6	297903.4
49	Zurich	US	Montana	-109.03	48.58	3120301202200	7760.7	305664.1
50	Paris	France		2.33	48.87	1133320101111	7336.2	313000.3
51	Berlin	Germany		13.37	52.50	1130131231000	874.3	313874.6
52	Boston	England		-0.02	52.98	4133123321322	901.0	314775.7
53	Zurich	Netherlands		5.38	53.10	1133102031320	360.9	315136.6
54	Dutton	England		-2.63	53.32	4133103103210	533.3	315669.9
55	Johannisburg	Germany		21.82	53.62	1130303323221	1609.5	317279.3
56	Johannisburg	Poland		21.82	53.63	1130303323201	1.1	317280.4
57	Moscow	Russia		37.58	55.75	1101302200021	1036.7	318317.2
58	Stockholm	Sweden		18.05	59.33	1132313023233	1225.4	319542.6
59	St.Petersburg	Russia		30.25	59.92	1102301202022	687.6	320230.1
60	Fairbanks	US	Alaska	-147.72	64.85	3102101033201	6135.6	326365.8

Table E-4						MEAN KM:	3590.7	
Spatial Ordering Statistics for 60 Cities and Towns						STD. DEV. KM:	3490.6	
Sorted by:	LONGITUDE					MIN & MAX KM:	1.1	15381.1
ORD	CITY	COUNTRY	STATE/PROV	LON DD	LAT DD	QTM ID	DIST LAST	DIST CUM
1	Fairbanks	US	Alaska	-147.72	64.85	3102101033201	12236.5	12236.5
2	Los Angeles	US	California	-118.25	34.07	3020233012200	3950.2	16186.7
3	Santiago	Mexico		-109.72	23.47	3323210132022	1439.9	17626.6
4	Zurich	US	Montana	-109.03	48.58	3120301202200	2790.7	20417.3
5	Santiago	Mexico		-99.85	25.42	3321213111303	2694.5	23111.8
6	Zurich	US	Kansas	-99.43	39.23	3322033301212	1534.9	24646.8
7	Los Angeles	US	Texas	-99.00	28.47	3320011020222	1196.2	25843.0
8	Paris	US	Texas	-95.55	33.67	3320321122013	664.4	26507.4
9	Bombay	US	Minnesota	-92.98	44.27	3022220022010	1198.3	27705.8
10	Jerusalem	US	Arkansas	-92.82	35.40	3322300303130	985.6	28691.4
11	Cairo	US	Illinois	-88.82	37.00	4322312021122	400.2	29091.6
12	Moscow	US	Mississippi	-88.65	32.77	4321333120230	470.3	29561.9
13	Santiago	Costa Rica		-84.30	9.85	4330210322333	2585.4	32147.3
14	Rome	US	Ohio	-82.42	38.47	4322030320121	3185.5	35332.8
15	Fairbanks	US	Florida	-82.27	29.73	4320310200030	971.2	36304.0
16	Zurich	CAN	Ontario	-81.62	43.43	4022202111312	1523.3	37827.3
17	Santiago	Panama		-80.98	8.10	4330003303103	3926.0	41753.4
18	St.Petersburg	US	Pennsylvania	-79.65	41.17	4022232132221	3676.8	45430.2
19	Wellington	CAN	Ontario	-77.35	43.95	4022011232332	361.7	45791.9
20	Dutton	US	Virginia	-76.47	37.50	4022132001302	720.5	46512.3
21	Santiago	Cuba		-75.82	20.02	4301223332310	1943.2	48455.6
22	Santiago	Peru		-75.73	-14.18	8300312130230	3800.0	52255.6
23	Los Angeles	Chile		-72.35	-37.47	8022103213202	2609.2	54864.8
24	Boston	US	Massachusetts	-71.07	42.37	4022301203222	8872.1	63736.9
25	Santiago	Dominican Rep.		-70.70	19.45	4303231303213	2546.9	66283.8
26	Santiago	Chile		-70.67	-33.45	8023111023230	5877.8	72161.6
27	Athens	US	Maine	-69.67	44.95	4022333113320	8711.7	80873.3
28	Stockholm	US	Maine	-68.13	47.05	4120330022123	261.9	81135.1
29	Sydney	CAN	Nova Scotia	-60.18	46.15	4121320200031	614.8	81750.0
30	Santiago	Bolivia		-59.57	-18.32	8313300201112	7163.6	88913.5
31	Santiago	Paraguay		-56.78	-27.15	8003311032233	1021.8	89935.3
32	Santiago	Brazil		-54.88	-29.18	8003210103222	292.4	90227.8
33	Colombo	Brazil		-49.23	-25.28	8003101331202	706.5	90934.2
34	Lagos	Portugal		-8.67	37.10	4233021133111	8124.8	99059.0
35	Dutton	England		-2.63	53.32	4133103103210	1861.6	100920.6
36	Boston	England		-0.02	52.98	4133123321322	178.0	101098.5
37	Paris	France		2.33	48.87	1133320101111	485.4	101583.9
38	Lagos	Nigeria		3.40	6.45	1222323202312	4714.4	106298.3
39	Zurich	Netherlands		5.38	53.10	1133102031320	5186.5	111484.7
40	Zürich	Switzerland		8.53	47.38	1133013131311	673.7	112158.4
41	Rome	Italy		12.48	41.90	1033003121132	684.1	112842.5
42	Berlin	Germany		13.37	52.50	1130131231000	1179.7	114022.1
43	Lagos	Angola		17.05	-16.07	5200013122120	7627.1	121649.3
44	Stockholm	Sweden		18.05	59.33	1132313023233	8378.3	130027.5
45	Wellington	South Africa		18.95	-33.63	5032111130232	10329.2	140356.8
46	Johannisburg	Germany		21.82	53.62	1130303323221	9698.4	150055.2
47	Johannisburg	Poland		21.82	53.63	1130303323201	1.1	150056.3
48	Athens	Greece		23.72	37.97	1030030332222	4866.9	154923.1
49	Berlin	South Africa		27.58	-32.90	5032022030212	7884.6	162807.7
50	Johannesburg	South Africa		28.00	-26.25	5032201301200	740.0	163547.7
51	St.Petersburg	Russia		30.25	59.92	1102301202022	9576.7	173124.4
52	Cairo	Egypt		31.25	30.05	1032313200120	3319.7	176444.1
53	Jerusalem	Israel		35.23	31.77	1002332101002	5112.9	181557.0
54	Moscow	Russia		37.58	55.75	1101302200021	2670.7	184227.8
55	Bombay	India		72.83	18.97	1300201313222	5033.7	189261.5
56	Colombo	Sri Lanka		79.85	6.93	1330020130001	1537.9	190799.3
57	Boston	Philippines		126.37	-7.87	6310213030209	5410.1	196209.5
58	Wellington	Australia		148.95	-32.55	6032301221221	3596.2	199805.7
59	Sydney	Australia		151.22	-33.87	6030322332302	257.0	200062.6
60	Wellington	New Zealand		174.78	-41.30	6233321230222	15381.1	215443.8

Table E-5						MEAN KM:	1450.6	
Spatial Ordering Statistics for 60 Cities and Towns						STD. DEV. KM:	1788.3	
Sorted by:	SHORTEST PATH					MIN & MAX KM:	1.1	10070.9
ORD	CITY	COUNTRY	STATE/PROV	LON DD	LAT DD	QTM ID	DIST LAST	DIST CUM
1	Athens	Greece		23.72	37.97	4022333113320	1051.37	1051.4
2	Rome	Italy		12.48	41.90	1130131231000	684.07	1735.4
3	Zürich	Switzerland		8.53	47.38	5032022030212	488.56	2224.0
4	Paris	France		2.33	48.87	1300201313222	485.35	2709.4
5	Boston	England		-0.02	52.98	3022220022010	177.96	2887.3
6	Dutton	England		-2.63	53.32	4133123321322	533.29	3420.6
7	Zurich	Netherlands		5.38	53.10	6310213030303	540.58	3961.2
8	Berlin	Germany		13.37	52.50	4022301203222	577.44	4538.6
9	Johannisburg	Germany		21.82	53.62	1032313200120	1.11	4539.7
10	Johannisburg	Poland		21.82	53.63	4322312021122	673.99	5213.7
11	Stockholm	Sweden		18.05	59.33	8003101331202	687.58	5901.3
12	St.Petersburg	Russia		30.25	59.92	1330020130001	633.85	6535.2
13	Moscow	Russia		37.58	55.75	4133103103210	2670.74	9205.9
14	Jerusalem	Israel		35.23	31.77	4022132001302	424.78	9630.7
15	Cairo	Egypt		31.25	30.05	3102101033201	3749.19	13379.9
16	Lagos	Portugal		-8.67	37.10	4320310200030	3618.39	16998.3
17	Lagos	Nigeria		3.40	6.45	1002332101002	2918.05	19916.3
18	Lagos	Angola		17.05	-16.07	3322300303130	1600.43	21516.7
19	Johannesburg	South Africa		28.00	-26.25	5032201301200	740	22256.7
20	Berlin	South Africa		27.58	-32.90	1130303323221	805.62	23062.4
21	Wellington	South Africa		18.95	-33.63	1130303323201	6545.8	29608.2
22	Colombo	Brazil		-49.23	-25.28	5200013122120	706.47	30314.6
23	Santiago	Brazil		-54.88	-29.18	1222323202312	292.41	30607.0
24	Santiago	Paraguay		-56.78	-27.15	4233021133111	1021.8	31628.8
25	Santiago	Bolivia		-59.57	-18.32	8022103213202	1783.25	33412.1
26	Santiago	Peru		-75.73	-14.18	3020233012200	2201.12	35613.2
27	Santiago	Chile		-70.67	-33.45	3320011020222	471.81	36085.0
28	Los Angeles	Chile		-72.35	-37.47	1101302200021	5142.17	41227.2
29	Santiago	Panama		-80.98	8.10	4321333120230	412.99	41640.2
30	Santiago	Costa Rica		-84.30	9.85	1133320101111	1450.16	43090.3
31	Santiago	Cuba		-75.82	20.02	3320321122013	539.18	43629.5
32	Santiago	Dominican Rep.		-70.70	19.45	1033003121132	1632.29	45261.8
33	Fairbanks	US	Florida	-82.27	29.73	4322030320121	693.61	45955.4
34	Moscow	US	Mississippi	-88.65	32.77	8313300201112	470.26	46425.7
35	Cairo	US	Illinois	-88.82	37.00	8003210103222	400.24	46825.9
36	Jerusalem	US	Arkansas	-92.82	35.40	8023111023230	315.24	47141.2
37	Paris	US	Texas	-95.55	33.67	4330210322333	664.43	47805.6
38	Los Angeles	US	Texas	-99.00	28.47	4301223332310	349.19	48154.8
39	Santiago	Mexico		-99.85	25.42	4303231303213	1021.32	49176.1
40	Santiago	Mexico		-109.72	23.47	3321213111303	1439.9	50616.0
41	Los Angeles	US	California	-118.25	34.07	3323210132022	1768.96	52385.0
42	Zurich	US	Kansas	-99.43	39.23	4330003303103	773.77	53158.7
43	Bombay	US	Minnesota	-92.98	44.27	8003311032233	914.29	54073.0
44	Zurich	CAN	Ontario	-81.62	43.43	8300312130230	298.75	54371.8
45	St.Petersburg	US	Pennsylvania	-79.65	41.17	1102301202022	361.69	54733.5
46	Wellington	CAN	Ontario	-77.35	43.95	4022232132221	538.25	55271.7
47	Boston	US	Massachusetts	-71.07	42.37	1132313023233	307.95	55579.7
48	Athens	US	Maine	-69.67	44.95	4120330022123	261.85	55841.5
49	Stockholm	US	Maine	-68.13	47.05	6030322332302	614.83	56456.3
50	Sydney	CAN	Nova Scotia	-60.18	46.15	4121320200031	1650.97	58107.3
51	Dutton	US	Virginia	-76.47	37.50	6032301221221	531.99	58639.3
52	Rome	US	Ohio	-82.42	38.47	4022011232332	2402.92	61042.2
53	Zurich	US	Montana	-109.03	48.58	6233321230222	2896.81	63939.0
54	Fairbanks	US	Alaska	-147.72	64.85	5032111130232	10070.86	74009.9
55	Bombay	India		72.83	18.97	4022202111312	1537.85	75547.7
56	Colombo	Sri Lanka		79.85	6.93	1133102031320	5410.11	80957.8
57	Boston	Philippines		126.37	-7.87	3322033301212	3596.2	84554.0
58	Wellington	Australia		148.95	-32.55	3120301202200	256.98	84811.0
59	Sydney	Australia		151.22	-33.87	1133013131311	2223.82	87034.8
60	Wellington	New Zealand		174.78	-41.30	1030030332222	2223.8	89258.7

Table E-6							MEAN KM:	10711.6	
Spatial Ordering Statistics for 60 Cities and Towns							STD. DEV. KM:	5428.2	
Sorted by:	LONGEST PATH						MIN & MAX KM:	298.8	19446.9
ORD	CITY	COUNTRY	STATE/PROV	LON DD	LAT DD	QTM ID	DIST LAST	DIST CUM
1	St.Petersburg	US	Pennsylvania	-79.65	41.17	4022232132221	298.8	298.8
2	Athens	Greece		23.72	37.97	1030030332222	17507.94	17806.7
3	Wellington	New Zealand		174.78	-41.30	6233321230222	19446.92	37253.6
4	Lagos	Portugal		-8.67	37.10	4233021133111	18149.22	55402.8
5	Sydney	Australia		151.22	-33.87	6030322332302	17026.39	72429.2
6	Sydney	CAN	Nova Scotia	-60.18	46.15	4121320200031	17101.06	89530.3
7	Wellington	Australia		148.95	-32.55	6032301221221	16773.96	106304.2
8	Dutton	England		-2.63	53.32	4133103103210	13203.19	119507.4
9	Boston	Philippines		126.37	-7.87	6310213030303	17755.64	137263.1
10	Santiago	Dominican Rep.		-70.70	19.45	4303231303213	15644.16	152907.2
11	Colombo	Sri Lanka		79.85	6.93	1330020130001	17440.04	170347.3
12	Santiago	Costa Rica		-84.30	9.85	4330210322333	15934.14	186281.4
13	Bombay	India		72.83	18.97	1300201313222	16614.49	202895.9
14	Santiago	Peru		-75.73	-14.18	8300312130230	12786.3	215682.2
15	Jerusalem	Israel		35.23	31.77	1002332101002	13511.98	229194.2
16	Los Angeles	Chile		-72.35	-37.47	8022103213202	14547.59	243741.8
17	Moscow	Russia		37.58	55.75	1101302200021	14118.02	257859.8
18	Santiago	Chile		-70.67	-33.45	8023111023230	13754.8	271614.6
19	St.Petersburg	Russia		30.25	59.92	1102301202022	12514.21	284128.8
20	Santiago	Brazil		-54.88	-29.18	8003210103222	13040.99	297169.8
21	Fairbanks	US	Alaska	-147.72	64.85	3102101033201	16435.59	313605.4
22	Berlin	South Africa		27.58	-32.90	5032022030212	16845.54	330450.9
23	Los Angeles	US	California	-118.25	34.07	3020233012200	16659.91	347110.8
24	Johannesburg	South Africa		28.00	-26.25	5032201301200	15744.87	362855.7
25	Santiago	Mexico		-99.85	25.42	3321213111303	14916.64	377772.3
26	Wellington	South Africa		18.95	-33.63	5032111130232	15440.36	393212.7
27	Zurich	US	Montana	-109.03	48.58	3120301202200	13954.29	407167.0
28	Lagos	Angola		17.05	-16.07	5200031322120	13418.1	420585.1
29	Santiago	Mexico		-109.72	23.47	3323210132022	11932.92	432518.0
30	Cairo	Egypt		31.25	30.05	1032313200120	11632.62	444150.6
31	Santiago	Panama		-80.98	8.10	4330003303103	10106.05	454256.7
32	Johannisburg	Germany		21.82	53.62	1130303323221	11694.67	465951.4
33	Santiago	Paraguay		-56.78	-27.15	8003311032233	11765.06	477716.4
34	Stockholm	Sweden		18.05	59.33	1132313023233	11211.55	488928.0
35	Colombo	Brazil		-49.23	-25.28	8003101331202	11085.77	500013.7
36	Johannisburg	Poland		21.82	53.63	1130303323201	11079.9	511093.6
37	Santiago	Bolivia		-59.57	-18.32	8313300201112	10508.74	521602.4
38	Berlin	Germany		13.37	52.50	1130131231000	8883.24	530485.6
39	Los Angeles	US	Texas	-99.00	28.47	3320011020222	10855.78	541341.4
40	Lagos	Nigeria		3.40	6.45	1222323202312	10636.89	551978.3
41	Zurich	US	Kansas	-99.43	39.23	3322033301212	8671.17	560649.5
42	Rome	Italy		12.48	41.90	1033003121132	8857.36	569506.8
43	Paris	US	Texas	-95.55	33.67	3320321122013	8253.76	577760.6
44	Zürich	Switzerland		8.53	47.38	1133013131311	7962.88	585723.5
45	Santiago	Cuba		-75.82	20.02	4301223332310	7655.02	593378.5
46	Zurich	Netherlands		5.38	53.10	1133102031320	7425.72	600804.2
47	Jerusalem	US	Arkansas	-92.82	35.40	3322300303130	7461.96	608266.2
48	Paris	France		2.33	48.87	1133320101111	7392.47	615658.6
49	Moscow	US	Mississippi	-88.65	32.77	4321333120230	7069.29	622727.9
50	Boston	England		-0.02	52.98	4133123321322	6910.62	629638.5
51	Fairbanks	US	Florida	-82.27	29.73	4320310200030	2276.72	631915.3
52	Stockholm	US	Maine	-68.13	47.05	4120330022123	2030.59	633945.8
53	Cairo	US	Illinois	-88.82	37.00	4322312021122	1826.77	635772.6
54	Athens	US	Maine	-69.67	44.95	4022333113320	1839.02	637611.6
55	Bombay	US	Minnesota	-92.98	44.27	3022220022010	1778.3	639389.9
56	Boston	US	Massachusetts	-71.07	42.37	4022301203222	1052.22	640442.2
57	Rome	US	Ohio	-82.42	38.47	4322030320121	741.52	641183.7
58	Wellington	CAN	Ontario	-77.35	43.95	4022011232332	720.47	641904.1
59	Dutton	US	Virginia	-76.47	37.50	4022132001302	789.35	642693.5
60	Zurich	CAN	Ontario	-81.62	43.43	4022202111312	298.75	642992.2

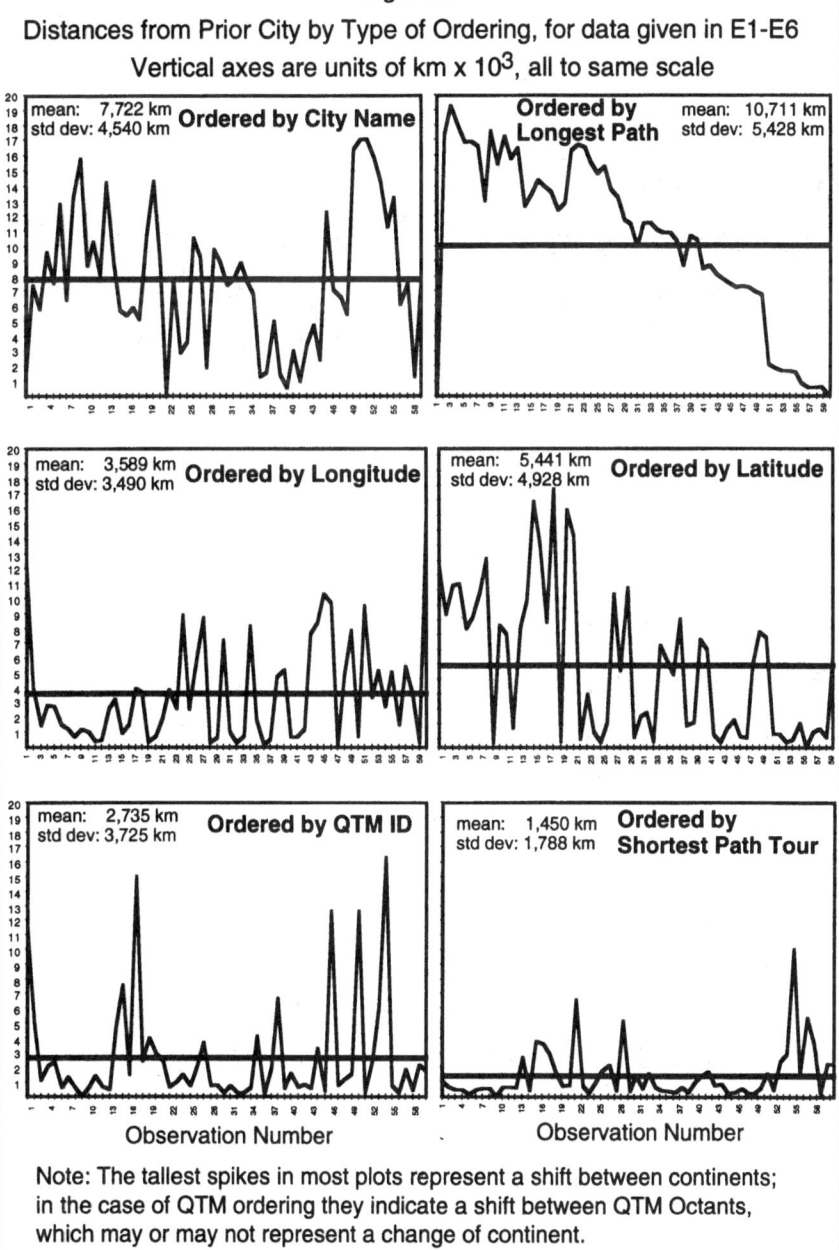

Figure E-7

Distances from Prior City by Type of Ordering, for data given in E1-E6
Vertical axes are units of km x 10^3, all to same scale

mean: 7,722 km **Ordered by City Name**
std dev: 4,540 km

Ordered by Longest Path mean: 10,711 km
std dev: 5,428 km

mean: 3,589 km **Ordered by Longitude**
std dev: 3,490 km

mean: 5,441 km **Ordered by Latitude**
std dev: 4,928 km

mean: 2,735 km **Ordered by QTM ID**
std dev: 3,725 km

mean: 1,450 km **Ordered by Shortest Path Tour**
std dev: 1,788 km

Observation Number

Observation Number

Note: The tallest spikes in most plots represent a shift between continents;
in the case of QTM ordering they indicate a shift between QTM Octants,
which may or may not represent a change of continent.

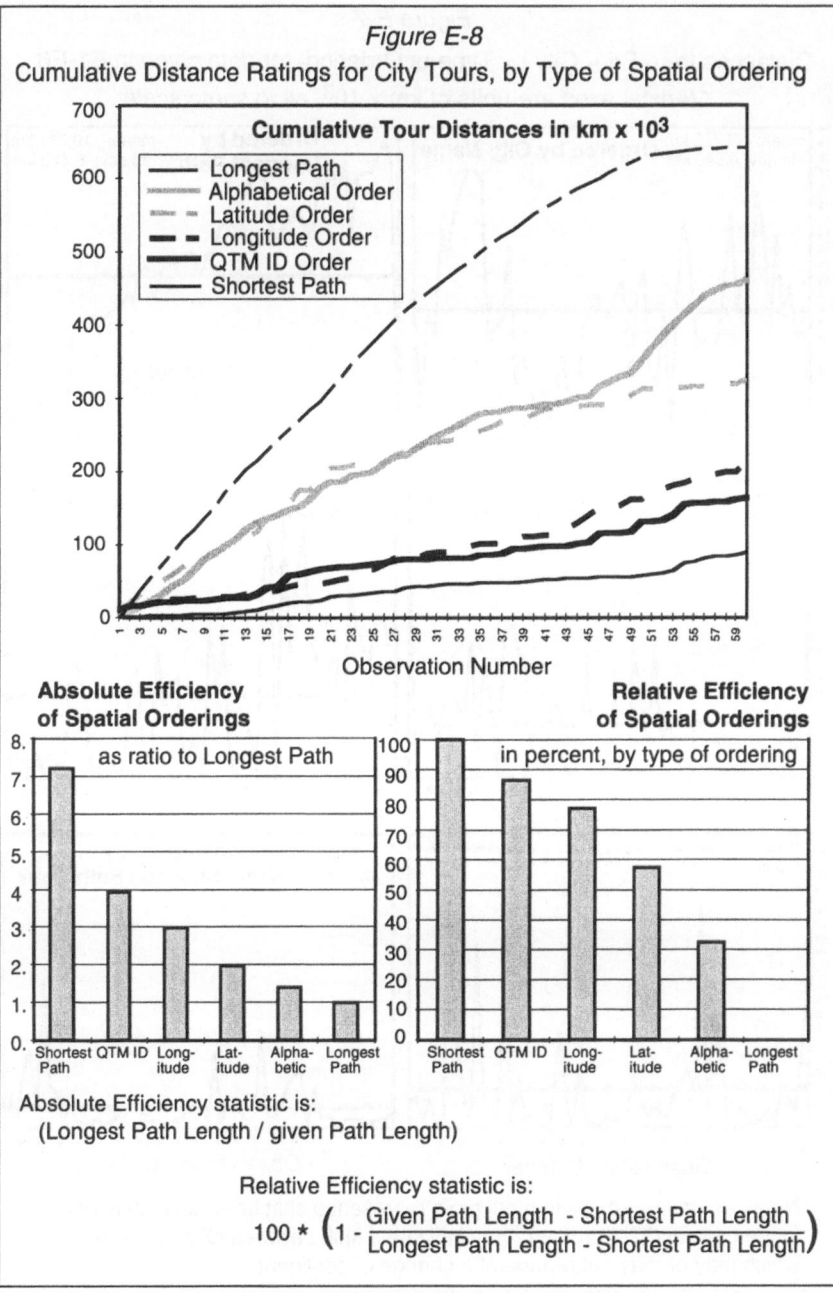

Figure E-8

Cumulative Distance Ratings for City Tours, by Type of Spatial Ordering

Cumulative Tour Distances in km x 10³

Longest Path
Alphabetical Order
Latitude Order
Longitude Order
QTM ID Order
Shortest Path

Observation Number

Absolute Efficiency of Spatial Orderings

as ratio to Longest Path

Shortest Path | QTM ID | Long-itude | Lat-itude | Alpha-betic | Longest Path

Relative Efficiency of Spatial Orderings

in percent, by type of ordering

Shortest Path | QTM ID | Long-itude | Lat-itude | Alpha-betic | Longest Path

Absolute Efficiency statistic is:
(Longest Path Length / given Path Length)

Relative Efficiency statistic is:

$$100 * \left(1 - \frac{\text{Given Path Length} - \text{Shortest Path Length}}{\text{Longest Path Length} - \text{Shortest Path Length}}\right)$$

References

Attneave, F., 1954. Some informational aspects of visual perception. *Psychological Review*. 61(3): 183-193.

Ballard, D.H and Brown, C.M., 1982. *Computer Vision*. Englewood Cliffs NJ: PrenticeHall.

Barber, C., Cromley, R. and Andrle, R., 1995. Evaluating alternative line simplification strategies for multiple representations. *Cartography and Geographic Information Systems*, 22(4): 276-290.

Barrett, P., 1995. Application of the linear quadtree to astronomical databases. *Astronomical Data Analysis Software & Systems IV*, ASP Conf. Series, v. 77.

Beard, M.K., 1991. Theory of the cartographic line revisited: Implications for automated generalization. *Cartographica*. 28(4): 32-58.

Brassel, K. and Weibel, R., 1988. A review and framework of automated map generalization. *Int. J. of Geographic Information Systems* 2(3): 229-44.

Bruegger, B.P., 1994. *Spatial Theory for the Integration of Resolution-Limited Data*. Ph.D. Dissertation, University of Maine, Orono, August 1994. [ftp:// grouse.umesve.maine.edu/pub/SurveyEng/Thesis/ Phd/Bruegger1994.PS.Z]

Bulmer, M.G., 1967. *Principles of Statistics*. Cambridge: The MIT Press. 2nd ed.

Buttenfield, B.P., 1984. Line structure in graphic and geographic space. Unpublished Ph.D. Dissertation, Dept. of Geography, University of Washington.

Buttenfield, B.P., 1985. Treatment of the cartographic line. *Cartographica* 22(2):1-26.

Buttenfield B. P., 1986. Digital definitions of scale-dependent line structure. *Proc. Auto-Carto London*. London: Royal Inst. of Chartered Surveyors, 497-506.

Buttenfield, B. P., 1989. Scale-dependence and self-similarity in cartographic lines. *Cartographica* 26(2): 79-100.

Chen, Z-T., 1990. A quadtree guides fast spatial searches in triangular irregular network (TIN). *Proc. Spatial Data Handling Symp. 4*. Dept. of Geography, U. of Zürich, July 1990, vol. 2, 209-215.

Chrisman, N.R., Dougenik, J. and White, D. 1992. Lessons for the design of polygon overlay processing from the Odyssey Whirlpool Algorithm. *Proc. Spatial Data Handling Symp. 5*. Charleston, SC, August 1992.

Chui, C.K., 1992. *An Introduction to Wavelets*. San Diego, CA: Academic Press.

Clarke, K.C. and Mulcahy, K.A., 1995. Distortion on the Interrupted Modified Collignon Projection. *Proc GIS/LIS 95*, Nashville TN.

Coxeter, H.S.M., 1973. *Regular Polytopes* (3rd ed.). New York: Dover.

Cromley, R.G., 1991. Hierarchical Methodsof Line Simplification. *Cartography and Geographic InformationSystems*, 18(2): 125-131.

Cromley, R.G. and Campbell, G.M., 1992. Integrating quantitative and qualitative aspects of digital line simplification. *The Cartographic Journal*, 29(1): 25-30.

de Berg, M., van Kreveld and M., Schirra, S. 1995. A new approach to subdivision simplification. *ACSM/ASPRS Ann. Convention and Exposition*, v. 4 (Proc. Auto-Carto 12): 79-88.

De Floriani, L .and Puppo, E., 1995. Hierarchical triangulation for multi-resolution surface description. *ACM Transactions on Graphics*,14(4): 363-411.

DeVeau, T., 1985. Reducing the number of points in a plane curve representation. *Proc. Auto Carto 7*. Falls Church, VA: ACSM, 152-60.

Devogele, T., Trevisan, J. and Raynal, L., 1997. Building a Multi-ScaleDatabase with Scale-Transition Relationships. in Kraak, M.J. and Molenaar, M. (eds) *Advances in GIS Research II* (Proc. SDH7, Delft, The Netherlands). London: Taylor & Francis, 337-352 .

Douglas D.H., Peucker T.K., 1973. Algorithms for the reduction of the number of points required to represent a digitized line or its caricature. *The Canadian Cartographer* 10(2): 112-22.

Dutton, G. and Chan, K., 1983. Efficient encoding of gridded surfaces. *Proc. Lincoln Inst. Colloquium on Spatial Algroithms for Microcomputer-based Land Data Systems*. Cambridge, MA: Lincoln Inst. for Land Policy.

Dutton, G., 1984. Truth and its consequences in digital cartography. *Proc. ASP-ACSM 44th Ann. Meeting*. Falls Church, VA: ACSM, 273-281.

Dutton, G., 1984a. Geodesic modelling of planetary relief. *Cartographica*. Monograph 32-33, 21(2&3): 188-207.

Dutton, G., 1989. The fallacy of coordinates. *Multiple Representations: Scientific report for the specialist meeting*, B.P. Buttenfield and J.S. DeLotto (eds.). Santa Barbara: NCGIA Technical Paper 89-3: 44-48.

Dutton, G., 1989a Modelling locational uncertainty via hierarchical tessellation. *Accuracy of Spatial Databases*, M. Goodchild & S. Gopal (eds.). London: Taylor & Francis, 125-140.

Dutton, G. 1990. Locational properties of quaternary triangular meshes. *Proc. Spatial Data Handling Symp. 4*. Dept. of Geography, U. of Zurich (July 1990) 901-10.

Dutton, G., 1991. Polyhedral hierarchical tessellations: The shape of GIS to come. *Geo Info Systems*. 1(3): 49-55.

Dutton, G., 1991a. Zenethial OrthoTriangular Projection. *Proc. Auto Carto 10*. Falls Church, VA: ACSM, 77-95.

Dutton, G., 1992. Handling positional uncertainty in spatial databases. *Proc. Spatial Data Handling Symp. 5*. Charleston, SC, August 1992, v. 2, 460-469.

Dutton, G. and Buttenfield, B.P., 1993. Scale change via hierarchical coarsening: Cartographic properties of Quaternary Triangular Meshes. *Proc. 16th Int. Cartographic Conference*. Köln, Germany, May 1993, 847-862.

Dutton, G., 1993a. Pathways to sharable spatial databases. Proc. AC 11. Bethesda MD: ASPRS/ACSM, Nov. 1993, pp. 157-166.

Dutton, G., 1996. Improving locational specificity of map data: A multi-resolution, metadata-driven approach and notation. *Int. J. of GIS*. London: Taylor & Francis, 10(3): 253-268.

Dutton, G., 1996a. Encoding and handling geospatial data with hierarchical triangular meshes. in Kraak, M.J. and Molenaar, M. (eds) *Advances in GIS Research II* (Proc. SDH7, Delft, Holland. London: Taylor & Francis, 505-518.

Dutton, G., 1997. Digital map generalization using a hierarchical coordinate system *Proc. Auto Carto 13*. (Seattle, WA) Bethesda, MD: ACSM/ASPRS, 367-376.

ESRI, 1996. Spatial Database Engine (SDE). *White Paper*, Environmental Systems Research Institute, March 1996, 7 p. [available via internet at: http://www.esri.com/base/common/whitepapers/whitepapers.html#SDE]

Fekete, G., 1990. Rendering and managing spherical data with sphere quadtrees. *Proceedings Visualization '90* (First IEEE Conference on Visualization, San Francisco, CA, October 23-26, 1990). Los Alamitos CA: IEEE Computer Society Press.

Fisher, I. and Miller, O., 1944. *World Maps and Globes*. New York: Essential Press.

FGDC, 1994. *Content Standards for Digital Geospatial Metadata*. Reston,VA: U.S. Federal Geographic Data Committee, June 8, 1994. Available by anonymous FTP from: fgdc.er.usgs.gov.

Franklin, W.R., 1984. Cartographic errors symptomatic of underlying algebra problems. *Proc. Int. Symposium on Spatial Data Handling*. Zürich, August 1984, vol. 1, 190-208.

Fuller, R.B., 1982. *Synergetics: Explorations in the geometry of thinking*. New York: Macmillan, 2 vols.

Goodchild, M.F., 1978. Statistical aspects of the polygon overlay problem. *Harvard Papers on Geographic Information Systems*. Dutton, G. (ed.). Reading MA: Addison-Wesley, vol. 6.

Goodchild, M.F.1988. The issue of accuracy in global databases. *Building Databases for GlobalScience*. H. Mounsey (ed.). London: Taylor and Francis, 31-48 .

Goodchild, M.F. and Gopal, S. (eds.), 1989. *Accuracy of Spatial Databases*. London: Taylor & Francis.

Goodchild, M.F. ,1990. Spatial Information Science (keynote address). *Proceedings, Fourth Int. Symp. on Spatial Data Handling* (Zurich, July 1990), vol. 1, 3-14.

Goodchild, M.F., Yang Shiren and Dutton, G. ,1991. Spatial Data Representation and Basic Operations for a Triangular Hierarchical Data Structure. Santa Barbara, CA: NCGIA Tech. Paper 91-8.

Goodchild, M.F., 1991a. Issues of quality and uncertainty. In J.C. Muller, ed., *Advances in Cartography*. New York: Elsevier, 113-40.

Goodchild, M.F. and Yang Shiren 1992. A hierarchical data structure for global geographic information systems. *Computer Graphics, Vision and Image Processing*, 54(1): 31-44.

Goodchild, M.F., 1995. Attribute accuracy. *Elements of Spatial Data Quality*. S.C. Guptill and J.L. Morrison (eds.). New York: Elsevier: 59-80.

Goodchild, M.F. and Hunter, G.J., 1997. A simple positional accuracy measure for linear features. *Int. J. of GIS*. 11(3): 299-306. London: Taylor & Francis.

Gray, R.W., 1994. Fuller's Dymaxion Map. *Cartography and Geographic Information Systems*. 21(4): 243-246.

Gray, R.W., 1995. Exact Transformation Equations for Fuller's World Map. *Cartographica* 32(3):17-25.

Guptill, S.C. and Morrison, J.L. (eds.), 1995. *Elements of Spatial Data Quality*. New York: Elsevier.

Guttman, A., 1984. R-Trees: A dynamic index structure for spatial searching. *Proc. ACM SIGMOD*. 47-57.

Hunter, G.J. and Goodchild, M.F., 1996. Communicating uncertainty in spatial databases. *Transactions in GIS* 1(1): 13-24.

Jasinski, M.J.,1990. The comparison of complexity measures for cartographic lines. Santa Barbara, CA: NCGIA Tech. Paper 90-1, 73 pp.

Jenks, G.F., 1981. Lines, computers and human frailties. *Annals of the Association of American Geographers* 71(1): 1-10.

Jones, C.B. and Abraham, I.M., 1986. Design considerations for a scale-independent database, *Proc. Spatial Data Handling 2*, 384-399.

Jones, C.B., J.M. Ware and G.L. Bundy, 1992. Multiscale spatial modelling with triangulated surfaces. *Proc. Spatial Data Handling 5*, vol. 2, 612-621.

Jones, C B, Kidner, D.B. and Ware, M., 1994. The implicit triangular irregular network and multiscale databases, *The Computer Journal*. 37(1), 43-57.

Kainz, W., 1987. A classification of digital map data models. *Proc. Euro-Carto VI*. M. Konecny (ed.) Brno, April 1987. U. of Brno, Dept. of Geography, 105-113.

Li, Z. and Openshaw, S., 1992. Algorithms for automated line generalization based on a natural principle of objective generalization. *Int. J. of GIS* 6(5): 373-390.

Li, Z. and Openshaw, S., 1993. A natural principle for the objective generalization of digital maps. *Cartography and Geographic Information Systems* 20(1): 19-29.

Lugo, J.A. and Clarke, K.C., 1995. Implementation of triangulated quadtree sequencing for a global relief data structure. *Proc. Auto Carto 12.* ACSM/ASPRS, 147-156.

Lukatella, H., 1989. Hipparchus data structure: Points, lines and regions in a spherical voronoi grid. *Proc. Ninth Int. Symp. on Computer-Assisted Cartography* (Auto-Carto 9), Baltimore, Maryland, 164-170.

Mark, D. M. and Lauzon, JP., 1985. The space efficiency of quadtrees, with emphasis on the effects of two-dimensional run-encoding. *Geo- Processing*, 2: 367-383.

Mark, D. M., 1989. Conceptual basis for geographic line generalization. *Proc. Ninth Int. Symp. on Computer-Assisted Cartography* (Auto-Carto 9), Baltimore, Maryland, 68-77.

Mark, D.M., 1991. Object modelling and phenomena-based generalization. *Map Generalization: Making Rules for Knowledge Representation*, Buttenfield, B.P. snd McMaster, R.B. (eds). London: Longman, 86-102.

Mandelbrot, B.B., 1982. *The Fractal Geometry of Nature.* San Francisco: Freeman.

Marino, J.S., 1979. Identification of characteristic points along naturally occuring lines: An empirical study. *The Canadian Cartographer* 16: 70-80.

McMaster, R.B., 1983. A mathematical evaluation of simplification algorithms. *Proc. Sixth Int. Symp. on Computer-Assisted Cartography* (Auto-Carto Six), Ottawa, Canada, 267-276.

McMaster, R.B., 1987. Automated Line Generalization. *Cartographica* 24(2):74-111.

McMaster, R.B., 1989 (ed). Numerical Generalization in Cartography. *Cartographica* 26(1).

McMaster, R.B., 1991. Conceptual frameworks for geographical knowledge. *Map Generalization: Making Rules for Knowledge Representation*, Buttenfield, B.P. snd McMaster, R.B. (eds). London: Longman, 21-39.

McMaster, R.B. and Shea K.S., 1992. *Generalization in DigitalCartography.* Washington, DC: Association of American Geographers, 134 p.

Moellering, H. and Rayner, J.N., 1982. The dual axis Fourier shape analysis of closed cartographic forms. *The Cartographic Journal* 19: 53-59.

Müller, J-C., 1986. Fractal dimension and consistencies in cartographic line representations. *The Cartographic Journal* 23: 123-30.

Müller, J.-C., 1987. Optimum point density and compaction rates for the representation of geographic lines. *Proc. Eighth Int. Symp. on Computer-Assisted Cartography* (Auto-Carto 8), Baltimore, Maryland, 221-230.

Müller, J.-C., Lagrange, J.-P. and Weibel, R. (eds.), 1995. *GIS and Generalization: Methodological andPractical Issues*. London: Taylor & Francis.

Nievergelt, J. and Widmayer, P., 1997. Spatial data structures: concepts and design choices. in Van Kreveld, M., Nievergelt, J., Roos, Th. and Widmayer, P. (eds.) *Algorithmic Foundations of Geographical Information Systems*. Lecture Notes in Computer Science, Berlin: Springer-Verlag, 153-197.

Nordenskiöld, A.E., 1889. *Facsimile-Atlas to the Early History of Cartography*. J.A. Ekelöf and C.R. Markham (trans.) New York: Dover, reprinted 1973.

ORACLE, 1995. Oracle7 Spatial Data Option: Advances in relational database technology for spatial data management. *Oracle White Paper* A30957, October 1995. [http://tiburon.us.oracle.com/odp/public/library/cr/pdf/27831.pdf]

van Oosterom, P., 1993. *Reactive Data Structures for Geographic Information Systems*. Oxford: Oxford University Press.

van Oosterom, P. and Schenkelaars, V., 1995. The development of a multi-scale GIS. *Int. J. of Geographic Information Systems* 9(5): 489-508.

Otoo, E.J. and Zhu, H., 1993. Indexing on spherical Surfaces using semi-quadcodes. *Advances in Spatial Databases*. Proc. 3rd Int. Symp. SSD'93, Singapore, 510-529.

Perkal, J., 1966. An attempt at objective generalization. *Michigan Inter-University Community of Mathematical Geographers*, Discussion Paper 10.

Peucker, T. 1976. A theory of the cartographic line. *Int. Yearbook of Cartography*, 16: 134-143.

Peuquet, D.J., 1984. A Conceptual Framework and Comparison of Spatial Data Models. *Cartographica*, 21(4): 66-113.

Plazanet C,, Affholder J-G, Fritsch E., 1995. The importance of geometric modeling in linear feature generalization. *Cartography and Geographic Information Systems* 22(4): 291-305.

Ramer, U., 1972. An iterative procedure for the polygonal approximation of plane curves. *Computer Graphics and Image Processing* 1: 244-56.

Richardson, L.F., 1961. The problem of contiguity: An appendix of statistics of deadly quarrels. *General Systems Yearbook* 6: 139-87.

Rieger, M. and Coulson, M., 1993. Concensus or confusion: Cartographers' knowledge of generalization. *Cartographica*. 30(1), 69-80.

Robinson, J.T., 1981. The K-D-B-Tree: A search structure for large multidimensional dynamic indexes. *Proc. ACM SIGMOD*, 10-18.

Samet, H. ,1990. *The Design and Analysis of Spatial Data Structures*. Reading MA: Addison-Wesley.

Ruas A., Plazanet C., 1996. Strategies for automated generalization. In Kraak M.j. and Molenaar, M. (eds.) *Advances in GIS research II* (Proc. SDH7, Delft, The Netherlands). London: Taylor & Francis, - .

Schirra, S., 1996. Precision and robustness. in Van Kreveld, M., Nievergelt, J., Roos, Th. and Widmayer, P. (eds.) *Algorithmic Foundations of Geographical Information Systems*. Lecture Notes in Computer Science, Berlin: Springer-Verlag, 255-287.

Schröder, P. and Sweldens, W., 1995. Spherical Wavlets: Efficiently representing functions on the sphere. *Computer Graphics* (SIGGRAPH '95 Proc.), New York: Association for Computing Machinery.

Snyder, J.P., 1987. Map Projections - A Working Manual. *US Geological Survey Prof. Paper 1395,* Washington: US Govt. Printing Office.

Snyder, J.P., 1992. An equal-area map projection for polyhedral globes. *Cartographica* 29(1): 10-21.

Timpf, S. and Frank, A.U., 1995. A Multi-Scale DAG for Cartographic Objects. *ACSM/ASPRS Annual Convention and Exposition*, (Proc.Auto-Carto 12). vol. 4, 157-163.

Tobler, W., 1988. Resolution, resampling and all that. *Building Databases for Global Science*. Mounsey, H and Tomlinson, R. (eds.). London: Taylor & Francis, 129-137.

Topfer, F. and Pillewizer, W., 1966. The principle of selection. *The Cartographic Journal* 3: 10-16.

Unknown ,1943. R. Buckminster Fuller's Dymaxion World. *Life*, March 1.

USDOC, 1992. *Spatial Data Transfer Standard* (Federal Information Processing Standard 173). Washington: Dept. of Commerce, National Institute of Standards and Technology.

USGS, 1990. *Digital line graphs from 1:24,000-scale maps--data users guide 1*: Reston, Virginia: U.S. Geological Survey.

USGS, 1994. *Standards for Digital Line Graphs*. Reston, Virginia: U.S. Geological Survey, 107 p.

Veregin, H., 1989. A Taxonomy of Error in Spatial Databases. Santa Barbara, CA: NCGIA Tech. Paper 89-12.

Visvalingam, M. and Whyatt , J.D.,1990. The Douglas-Peucker algorithm for line simplification: Re-evaluation through visualization. *Computer Graphics Forum* 9(3): 213-228.

Watson, D.F., 1988. Natural neighbor sorting on the n-dimensional sphere. *Pattern Recognition* 21(1): 63-67. [papers, source code and Java demo available at http://www.iinet.com.au/~watson/modemap.html]

Watson, D.F., 1994. *nngridr: An implementation of natural neighbor interpolation*. published by David Watson, P.O.Box 734, Claremont, WA 6010, Australia, 170p.

Waugh, T.C. 1986. A response to recent papers and articles on the use of quadtrees for geographic information systems *Proc. 2nd Int. Symp. on Spatial Data Handling*. (Seattle WA, July 1986), 33-37.

Weibel, R., 1991. Amplified intelligence and rule-based systems. *Map Generalization: Making Rules for Knowledge Representation*, Buttenfield, B.P. and McMaster, R.B. (eds). London: Longman, 172-186.

Weibel, R., 1995. Map Generalization. Special Issue, *Cartography and Geographic Information Systems* 22(4).

Weibel, R., 1997. A typology of constraints to line generalization. in Kraak, M.J. and Molenaar, M. (eds) *Advances in GIS Research II* (Proc. SDH7, Delft, The Netherlands). London: Taylor & Francis, 533-546.

Weibel, R. and Dutton, G. (in press). Generalization of spatial data and dealing with multiple representations, in Longley, P., Goodchild, M., Maguire, D., and Rhind, D. (eds.), *Geographical Information Systems: Principles, Techniques, Management and Applications*, v. 1. John Wiley.

White, D., Kimmerling, J. and Overton, W.S., 1992. Cartographic and geometric components of a global sampling design for environmental monitoring. *Cartography and Geographic Information Systems*. 19(1): 5-22.

White, E.R., 1985. Assessment of line-generalization algorithms using characteristic points. *The American Cartographer* 12(1): 17-27.

Wickman, F.P., Elvers, E. and Edvarson, K., 1974. A system of domains for global sampling problems. *Geografiska Annaler*, 56A 3-4, 201-211.

Zhan, F.B. and Buttenfield, B.P., 1996. Multi-scale representation of a digital line. *Cartography and Geographic Information Systems*. 23(4): 206-228.

Zhao, Z. and Saalfeld, A. 1997. Linear-time sleeve-fitting polyline simplification algorithm. *ACSM/ASPRS Annual Convention and Exposition*, v. 5 (Proc. Auto-Carto 13): 214-223.

ZYCOR, Inc., 1984. Manual of automated feature displacement. *Report for U.S. Army Engineering Topographic Laboratories*, Fort Belvoir, VA, 204 p

Index

About the author

Geoffrey H. Dutton is a Research Associate in the Department of Geogrpahy, University of Zürich, Switzerland, where he received his Ph.D. in 1998, under the direction of Professors Robert Weibel and Kurt Brassel. Dr. Dutton has a B.A. degree in Government from Columbia University and a Masters in City Planning from Harvard University. From 1969 through 1984 he was a staff researcher at the Laboratory for Computer Graphics and Spatial Analysis at Harvard's Graduate School of Design, where he developed and applied cartographic and geographic modeling software, in both theoretical and project-oriented contexts. Subsequently he has been a private consultant for industrial, nonprofit and academic organizations, worked as a software designer in the computer industry, and has been senior GIS analyst at several design and mapping firms in the Boston area. Over the past 25 years Dr. Dutton has authored more than 50 professional papers, articles and book chapters, and contributed articles and maps to trade journals and news magazines. He is married and lives in Zürich and Somerville, Massachusetts.

Springer
and the
environment

At Springer we firmly believe that an
international science publisher has a
special obligation to the environment,
and our corporate policies consistently
reflect this conviction.
We also expect our business partners –
paper mills, printers, packaging
manufacturers, etc. – to commit
themselves to using materials and
production processes that do not harm
the environment. The paper in this
book is made from low- or no-chlorine
pulp and is acid free, in conformance
with international standards for paper
permanency.

 Springer

Lecture Notes in Earth Sciences

For information about Vols. 1–19
please contact your bookseller or Springer-Verlag